HANDBOOK OF INSTRUMENTATION
AND TECHNIQUES FOR SEMICONDUCTOR
NANOSTRUCTURE CHARACTERIZATION

World Scientific Series in Materials and Energy

Series Editor: Leonard C. Feldman *(Rutgers University)*

Vol. 1   Handbook of Instrumentation and Techniques for Semiconductor Nanostructure Characterization
edited by Richard Haight (IBM TJ Watson Research Center, USA),
Frances M. Ross (IBM TJ Watson Research Center, USA),
James B. Hannon (IBM TJ Watson Research Center, USA) and
Leonard C. Feldman (Rutgers University, USA)

Vol. 2   Handbook of Instrumentation and Techniques for Semiconductor Nanostructure Characterization
edited by Richard Haight (IBM TJ Watson Research Center, USA),
Frances M. Ross (IBM TJ Watson Research Center, USA),
James B. Hannon (IBM TJ Watson Research Center, USA) and
Leonard C. Feldman (Rutgers University, USA)

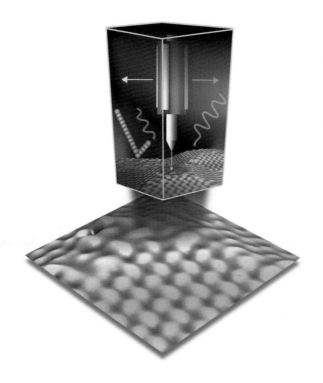

MATERIALS AND ENERGY – Vol. 2

HANDBOOK OF INSTRUMENTATION AND TECHNIQUES FOR SEMICONDUCTOR NANOSTRUCTURE CHARACTERIZATION

Editors

Richard Haight • Frances M Ross • James B Hannon

IBM TJ Watson Research Center, USA

with Foreword by Leonard C. Feldman

 **World Scientific**

NEW JERSEY · LONDON · SINGAPORE · BEIJING · SHANGHAI · HONG KONG · TAIPEI · CHENNAI

Published by

World Scientific Publishing Co. Pte. Ltd.

5 Toh Tuck Link, Singapore 596224

USA office: 27 Warren Street, Suite 401-402, Hackensack, NJ 07601

UK office: 57 Shelton Street, Covent Garden, London WC2H 9HE

British Library Cataloguing-in-Publication Data
A catalogue record for this book is available from the British Library.

World Scientific Series in Materials and Energy
HANDBOOK OF INSTRUMENTATION AND TECHNIQUES FOR SEMICONDUCTOR NANOSTRUCTURE CHARACTERIZATION
(In 2 Volumes)

ISBN-13 978-981-4322-80-5 (Set)
ISBN-10 981-4322-80-6   (Set)

ISBN-13 978-981-4322-81-2 (Vol. 1)
ISBN-10 981-4322-81-4   (Vol. 1)

ISBN-13 978-981-4322-82-9 (Vol. 2)
ISBN-10 981-4322-82-2   (Vol. 2)

Typeset by Stallion Press
Email: enquiries@stallionpress.com

Printed in Singapore by Mainland Press Pte Ltd.

CONTENTS

VOLUME 2

SCANNING PROBES

ATOM AND OPTICAL PROBES

FOREWORD

Modern semiconductor materials science is synonymous with atomic level artistry. The practitioners are artists; their palette is the periodic table. Scientists and engineers use their knowledge of atoms and the behavior of atoms in a solid to create new chemical combinations and new forms of materials with wonderful properties. World-class accomplishments in such materials have been recognized by Nobel prizes in Physics and Chemistry and given rise to entirely new technologies. Materials science advances have underpinned the semiconductor technology revolution that has driven societal changes for the last fifty years.

The end is not in sight! Future technology-based problems dominate the current scene. High on the list are control and conservation of energy and the environment, water purity and availability, and propagating the information revolution. All fall in the technology domain. In every case, solutions begin with new forms of materials, materials processing or new artificial material structures. Scientists seek new forms of semiconductor photovoltaics with greater efficiency and lower cost. Revolutionary concepts will extend the information revolution by controlling the spin of electrons or enabling quantum states as in quantum computing. It is this latter category, the semiconductor materials revolution that receives special attention in these volumes.

Crucial to all materials development is the ability to "see" the final product. About 100 years ago we realized that "seeing" at the atomic level meant bringing new forms of radiation into play — electromagnetic radiation far beyond the visible that interacted with solids as only waves can, and particle-induced quantum radiation to extend the wavelengths and spatial dimensions. This explosion in imaging capability has offshoots in the medical sciences, archeology, anthropology, environmental science and almost every scientific field, even forensics. Invariably the first developments arise from the physical sciences which provide the basis of understanding of the interaction of radiation with matter. Combining these basics with the creativity of scientists and the ingenuity of instrument makers has created an advanced set of instrumentation and techniques that allows the crucial goal of "seeing" the atoms — their type and their arrangement. This development has been on-going for the last century. Nevertheless the needs and the advances took an exponential step forward with the onset of nanotechnology. The challenge was to probe on the few atom scale and extract the quantitative information necessary to assure that the final "painting" indeed recreated the artist's original conception. This capability led to the inevitable cycle of semiconductor nano-materials development; namely materials creation through processing, materials characterization through nano-scale investigation, determination of electronic or optical properties, and the feed-back process driving to perfection. This book describes advances in the critical link of this cycle, namely the ability to carry out accurate semiconductor characterization at the nano-scale.

Chapters in this book, authored by prominent members of the community, describe the most recent advances in this critical step of the materials cycle. Progress has been remarkable. Volume 1 considers electron microscopies and x-ray diffraction techniques — both previously established techniques that have undergone astonishing transformations to meet the nano-materials challenge. Volume 2 considers the recently developed tools designated as proximal probes and sophisticated optical probes. The motivation is ever more sophisticated materials characterization. One cannot over emphasize the ingenuity and creativity of this aspect of the materials revolution. Indeed, advances in characterization have been recognized in their own right through Nobel prizes, society awards and broad public recognition. The author list includes individuals who have been so honored.

Finally it is appropriate to comment on the future. Progress in this field emerges from the ingeniousness of those who understand the processes and the needs of new semiconductor materials. Many such examples are contained in the following chapters. Each link in the cycle is critical; advances in one undoubtedly lift the entire process. As mentioned above, progress in materials characterization not only advances semiconductor technology but eventually spins-off to other scientific fields — all highly dependent on atomic scale imaging. So the latest aspects described here will have broad impact throughout the sciences, from biology to pharmacology to cite just a few.

Materials characterization science has established a remarkable record of accomplishment to date, responding to technological needs. Each generation of technology has been improved as the community addressed the materials limitations. The world we face of nano-scale manipulation and precise atomic control will be even more demanding and require new techniques capable of moving the ever expanding frontiers of semiconductor science and technology. Fortunately the field will probably never peak, but continue to break new ground, illuminating the path to still greater materials control and analysis. These volumes provide the very latest in this critical technology and are an invaluable resource for scientists in both academia and industry concerned with the semiconductor future and all of science.

L C Feldman
Institute of Advanced Materials, Devices and Nanotechnology
Rutgers University

April, 2011

INTRODUCTION

History has shown that scientific breakthroughs are often a result of the innovative invention, development and implementation of new scientific instruments and techniques. The discovery of the penetrating properties of X-rays not only brought a Nobel prize for Roentgen but also revolutionized the fields of medicine, biology, physics and materials science. The invention of the laser resulted in a Nobel for Townes, Basov and Prokhorov and led to additional prizes in physics and chemistry that spawned the fields of non-linear optics, laser cooling, precision laser spectroscopy, Bose-Einstein condensation and ultrafast spectroscopies that reach into the femtosecond and attosecond time regimes. In nanoscience, the invention and development of the electron microscope (Nobel for Ruska) in all of its variations, and more recently the scanning tunneling microscope (Nobel for Binnig and Rohrer) have driven a veritable revolution in our understanding of materials and their surfaces at the nanoscale. The list of key scientific advances linked to breakthroughs in instrumentation and techniques could fill an entire chapter of this Handbook.

Nowhere is the development of instrumentation and associated techniques more important to scientific progress than in the area of nanoscience. As we delve more deeply into the physics and chemistry of inorganic, organic and biological materials and processes, we are inexorably driven to the nanoscale. Even in the commercial realm, such as, for example, the technology of integrated circuit manufacturing, feature sizes already extend well into the nanoscale. The need to "see" and understand processes at the nanoscale has driven the development of entirely new instruments, stimulated advancement of older instruments in unexpected ways, and led to new techniques that often yield profound insights. Because of this explosion in scientific efforts to peer into materials and processes at the nanoscale, it is critical to develop a resource that captures many of the key instruments and techniques that are the modern day staple for scientific investigation in this arena. Semiconductor nanostructures provide a focal point for instrumentation development, since their applications require both precise control of structure and composition and accurate measurement of electronic and optical properties.

In this Handbook, we have assembled chapters written by eminent scientists who have advanced the forefront of instrumentation and developed key techniques for the study of structural, optical and electronic properties of semiconductor nanostructures. The handbook is organized into four main sections, **Electron Microscopies, X-Ray Diffraction Techniques, Scanning Probes,** and **Atom and Optical Probes.**

In **Volume 1**, within the section entitled **Electron Microscopies**, we begin with a detailed discussion of the ubiquitous scanning electron microscope in Chapter 1. The SEM is arguably one of the most widely used instruments for studying materials at the nanoscale, and SEMs in various forms can be found in most academic and industrial laboratories. In Chapter 2, we cover the use of transmission electron microscopy for structural measurements, and describe a more exotic technique, ultra-high vacuum transmission electron microscopy, that can provide unique information on growth mechanisms and defect formation within semiconductor nanostructures. Chapter 3 discusses aberration corrected scanning transmission electron microscopy. Aberration correction, analogous to that practiced in light optics, leads to dramatic improvement in the performance of electron microscopes, and its incorporation into commercial systems is critical to achieving the resolution required for analysis of nanostructures. In Chapter 4 we describe the state of the art in low energy electron microscopy (LEEM), a technique that provides highly detailed structural and electronic information on semiconductor surfaces. Both aberration correction and applications to photoelectron microscopy (PEEM) are discussed. Chapter 5 describes the unique nanoscale physics that occurs when light interacts with material surfaces, generating plasmons with nanoscale spatial dimensions and femtosecond dynamic behavior. This unique approach combines LEEM and PEEM with femtosecond light pulses to create an instrument and technique that merges two scientific frontiers of time and spatial resolution.

We next move on to **X-ray Diffraction Techniques**, where Chapter 6 provides a thorough discussion of the basics of X-ray diffraction and its application to the study of strain in epitaxial nanostructured systems, highlighting, for example capped nanostructures and chemical composition analysis by anomalous X-ray diffraction. Chapter 7 discusses the details of X-ray diffraction with emphasis on focusing of X-rays with refractive, reflective and capillary optical approaches. In these chapters the authors describe the exciting advances made possible by availability of ultrahigh brightness synchrotron sources.

In **Volume 2**, within the section entitled **Scanning Probes**, Chapters 8 and 9 review the basics of scanning probe microscopy and spectroscopy, and describe the most recent advances in scanning probe techniques that include atomic force microscopy and its variants (Kelvin probe, scanning capacitance microscopy, and scanning spreading resistance microscopy) and scanning tunneling microscopy and spectroscopy. The focus is on real-life applications of these techniques for the study in particular of III-V semiconductors such as arsenides and nitrides, two materials systems of key technological importance in a range of applications. Surface structures such as quantum dot sizes and shapes, interdiffusion, segregation and doping can be addressed using these techniques.

In the final section of the Handbook, entitled **Atom and Optical Probes**, we present four chapters describing experimental approaches that exploit the interaction of atoms or light to interrogate the state of nanostructured materials. Chapter 10 describes atom probe tomography, an approach that provides extraordinary spatial resolution combined with elemental sensitivity to study materials critical to the microelectronics industry at the nanoscale. Chapters 11 and 12 present two techniques showing the utility of optical probes at the nanoscale by isolating and studying individual nanostructures under focused beams of visible and deep ultraviolet light. In Chapter 11, the fundamentals of Raman spectroscopy are presented to illuminate electronic and vibrational physics at the nanoscale, with applications to the investigation of individual carbon nanotubes and graphene. In Chapter 12, a unique single nanowire photoelectron spectroscopy instrument is described. Basic photoelectron spectroscopy physics is applied to individual nanowires and nanotubes with applications to Si, Ge, carbon nanotubes and doping of semiconductor nanowires. Chapter 13 finishes out the Handbook with a discussion of thermoreflectance techniques for the study of thermal transport in nanostructures.

The chapters in this Handbook cover the most important cutting edge instrumentation and techniques available to the experimentalist studying semiconductors at the nanoscale. While the field of nanoscience will undoubtedly continue to advance at a breakneck pace, the material and references presented in this Handbook provide an extraordinarily useful base of information for thorough investigations of nanoscale semiconductor structures.

Richard Haight
Frances Ross
James Hannon
IBM T. J. Watson Research Center, Yorktown Heights., N.Y.

February, 2011

8

AN INTRODUCTION TO SCANNING PROBE MICROSCOPY OF SEMICONDUCTORS WITH CASE STUDIES CONCERNING GALLIUM NITRIDE AND RELATED MATERIALS

Rachel Oliver
Department of Materials Science and Metallurgy
University of Cambridge, Pembroke Street
Cambridge CB2 3QZ, UK

8.1. INTRODUCTION

The aim of this chapter is to introduce the use of scanning probe microscopy (SPM) techniques in the characterization of semiconductors at the micro- and nano-scale. Two key techniques will be introduced: scanning tunneling microscopy (STM) and atomic force microscopy (AFM). Additionally, a brief outline will be provided of some of the common SPM approaches used to elucidate the electrical properties of semiconductors. However, it would not be helpful to discuss microscopy in isolation from the examination of real materials. Hence, examples will be provided of the applications of widely-used SPM techniques drawn from the growing body of research on a recently-developed semiconductor material — gallium nitride (GaN) — and its alloys. Where possible, SPM data on GaN will be compared with data from other microscopy techniques (e.g. transmission electron microscopy (TEM) and atom probe tomography (APT)), to highlight the strengths and weaknesses of SPM and the ways in which complementary data from multiple approaches can aid our understanding of semiconductor materials. A brief introduction to GaN will be provided in this first section, to provide context for case studies in later sections on the application of SPM in nitride semiconductor science.

The main importance of GaN lies in its application in optoelectronic devices, such as light emitting diodes (LEDs) and laser diodes (LDs). GaN is a wide bandgap semiconductor, and with its alloys InGaN and AlGaN it is used to make devices emitting in the blue, green and ultra-violet spectral regions. More conventional compound semiconductor materials such as GaAs have narrower bandgaps, and cannot be used to directly access this spectral region. Unlike GaAs and other similar materials, GaN is very difficult to grow in a bulk form, and hence GaN-based devices are usually grown by either molecular beam epitaxy (MBE) or metal-organic vapor phase epitaxy (MOVPE) on substrates such as sapphire or silicon carbide. This heteroepitaxy results in a high density of crystal defects in these materials, with dislocation densities of 10^8–10^9 cm^{-2} being common in LED structures. (The assessment of dislocation densities is one very common application of SPM in the GaN context).[1] The successful operation of GaN-based optoelectronic devices in the presence of such large numbers of dislocations is rather surprising, since in more conventional compound semiconductors such high defect densities would be expected to cause very rapid device failure.[2] One contribution of AFM to the debate on how such dislocation-related device failure is avoided in the GaN system is described in Sec. 8.5.

GaN also differs from materials like GaAs in more fundamental ways. It has a wurtzite crystal structure, whilst GaAs and other arsenide semiconductors have the zincblende crystal structure. GaN also exhibits very high piezo-electric constants resulting in large built-in electric fields in strained heterostructures such as quantum wells. The difference in crystal structure between GaN and the more thoroughly-studied arsenides makes studies of the detailed surface atomic structures of GaN particular interesting, and a case study outlining the use of STM in this context is given in Sec. 8.4. These types of STM studies on clean semi-conductor surfaces under vacuum conditions are particularly relevant to the growth of nitrides by MBE.

Whilst the most common applications of GaN to date are in opto-electronics, it also has potential applications in high power electronic devices such as high electron mobility transistors (HEMTs). In this case control of the doping in the structure is vital, and can be challenging since most GaN is found to be unintentionally n-doped due to impu-rity incorporation. The final case study in this chapter will illustrate the application of a popular AFM-based electrical characterization tool — scanning capacitance microscopy — in quantifying unintentional doping and helping to identify why it occurs.

Before these various case studies however, the basic principals of the key SPM techniques will be described, along with some practical aspects of their operation.

8.2.   SCANNING TUNNELING MICROSCOPY

8.2.1.   Basic Principles of Scanning Tunneling Microscopy

In scanning tunneling microscopy (STM), a small metallic needle, called the *tip* is scanned over a sample surface. The tip and sample are sep-arated by a tiny, empty gap and a bias is applied between them. In a classical worldview, we would not expect any current to be able to flow across the gap. However, in a quantum worldview, the electron has both a wave and a particle nature. When an electron is incident upon a potential barrier (such as the vacuum gap) we may use the Schrödinger equation to find the electron's wavefunction, and show that this function penetrates into the vacuum gap.[3] This penetration of the wavefunction into the classically forbidden region represents a finite probability of finding the electron in the gap, and indeed, for a narrow gap, there is a finite probability of the electron crossing the gap. Hence, under the application of a bias, a current will flow across the gap. This current is known as the tunneling current, I, and for an STM the variation of I with the tip-sample distance d may be approximated using the following expression:

$$I \propto \exp(-2Kd) \qquad (8.1)$$

where,

$$K = \frac{1}{\hbar}(2m\Phi)^{\frac{1}{2}} \qquad (8.2)$$

and Φ is the average of the workfunctions of the tip and the sample.

This exponential dependence of the tunneling current on the tip-sample separation is what gives STM its power. A very small change in d leads to a very big change in the current. For a typical value of Φ (4 eV), K is about 1Å^{-1}. This implies that a 1Å change in the tip-sample separation will lead to a change in the tunneling current of about an order of magnitude. Hence, if the apex of the tip terminates in a single atom, then the vast majority of the tunneling current will flow between that atom and the nearest atom on the surface. Atoms a little way away from the apex of the tip will barely contribute to the tunneling current. Thus, by scanning the tip over the surface in x and y, we might expect to produce an image in which the current varies as we scan over atomic protuberances on the surface.

STM can indeed be used to take atomic resolution images. However, once we start to think about the tunneling current flowing between the tip apex atom and a nearby surface atom, our simple equation for tunneling becomes hard to understand. Unfortunately, it is not obvious how you define the shortest distance between two atoms! You would have to start to think about the shape of the electron cloud which surrounds the nucleus of the atom. Not long after the invention of STM by Binnig *et al.*,[4] Tersoff and Hamann[5] attempted to produce a model which did just that. They modeled the apex of the tip as a small sphere, and hence treated its electronic structure as consisting only of spherically-symmetrical *s*-wave functions. Other assumptions of their model include that the experiment is performed at low temperature and low voltage. Within these assumptions, they conclude that the tunneling current is proportional to the local density of states (LDOS) of the sample surface at the Fermi level (E_F). Hence, if we consider a notional STM image, in which the tip is scanned across the surface and a constant tunneling current is maintained, we would map out a contour at which the surface

LDOS at the Fermi level is constant. This means that a simple picture of STM imaging, which suggests that STM measures the *topography* (i.e. the variation in height) of the surface is incomplete.[6] An STM image will also be affected by electronic effects. The Tersoff and Hamann model[5] of STM, emphasizing the role of the LDOS at the sample surface, underpins current understanding of STM.

Whilst in this short introduction, there is insufficient space to detail the extensive developments that have occurred in the modeling of STM since these seminal studies of the early 1980s, it is worth noting that the Tersoff and Hamann model[5] has some shortcomings, which its authors acknowledged. The assumptions of low temperature and low voltage do not hold for most studies on semiconductors, which may exhibit insufficient conductivity at low temperatures for any STM imaging to be performed, and which equally may require moderate voltages to be applied before a measurable tunneling current is observed. Additionally, whilst the Tersoff and Hamann model[5] provides good agreement with experimental studies of gold surfaces, experimental studies on other materials suggest that the resolution of STM is better than that which the model would predict. This may be because, since the tip is modeled as a pure *s*-wave, it does not incorporate the geometry or electronic structure of a real tip. Real tips are typically made from *d*-band metals such as Pt, Ir and W, and further modeling work by Chen[7] suggested that the highly localized d_{z^2} dangling bonds may be responsible for the excellent resolution exhibited in STM images of materials such as Al and Cu. For further details on the development of models for STM, the reader is referred to the extensive review by Hofer *et al.*[8]

Within the picture of STM imaging as accessing the LDOS of the surface, we can consider the impact of imaging at different voltages. When the sample is negatively biased relative to the tip, electrons will tunnel from the occupied states of the sample to the unoccupied states of the tip. If the sample is positively biased, electrons will tunnel from the occupied states of tip to the unoccupied states of the sample. This effect provides useful information about the electronic structure of the surface, and is particularly interesting in the imaging of compound semiconductors such as GaAs. The strong electronegativity difference between Ga and As results in the As sites in this compound being rather electron rich, whilst the Ga sites are depleted of electrons. Hence, if we have a negatively biased sample, we will primarily image the occupied electron states on the As sites, whereas in an unoccupied states image, tunneling will primarily occur into the Ga sites. For imaging of a semiconductor alloy such as InGaAs, this provides a convenient way of separating topographic and chemical effects. In imaging the non-metal sublattice, any changes in contrast between InAs and GaAs will be largely related to topography (although of course the image will still be influenced by the electronic structure around the As atoms). In imaging the metallic sub-lattice, the different metal species (In and Ga) will result in differences in the LDOS, so that in addition to topographic contrast there will be chemical information present.[9]

Going beyond imaging at different bias polarities, it is possible to use the STM spectroscopically. Curves of current (I) versus voltage (V) may be recorded at specific sites, with the tip height constant. This technique is known as scanning tunneling spectroscopy (STS). Whilst interpretation of this data is complex and requires some care,[6] for biasing conditions in which electrons are injected from the occupied states of the tip into the unoccupied states of the sample, peaks on a plot of dI/dV against V, can be interpreted as corresponding to energies with a high LDOS.[6] This approach can be taken beyond recording site-specific I-V curves into the realms of spectrum imaging — i.e. a spectrum can be taken at every pixel of an image forming a data cube. However, it is important to ensure in this case that topographic and electronic effects do not become mixed, and various techniques which allow imaging of both the surface topography and dI/dV versus V curves have been developed, such as current imaging tunneling spectroscopy (CITS).[10] A useful overview of the challenges of STS and related techniques is given by Kubby and Boland.[6]

8.2.2. Some Practical Considerations in STM

Having provided a brief overview of the physical principals of STM, it is worth mentioning some practical issues. This chapter will not provide an extensive description of STM instrumentation, however. For further details concerning the practicalities of building an STM system, the reader is once again referred to Kubby and Boland.[6]

The STM tip is scanned over the sample surface using a piezo-electric crystal. Piezo-electric materials expand or contract upon the application of a voltage and by applying carefully calibrated voltage ramps scanning in the x and y directions may be achieved. Section 8.2.1 described a notional STM image in which the tunneling current was kept constant and the tip mapped out a contour of constant LDOS at the surface. This type of imaging is known as *constant current imaging* and is achieved in practice by using a feedback circuit to control a voltage applied to the piezo-scanner in order to continuously adjust the tip height (i.e. the extension of the piezo-scanner in z) so that a setpoint tunneling current chosen by the user is maintained as far as possible. The variation in extension of the piezo-crystal with voltage is carefully calibrated so that the varying voltages applied by the feedback circuit during the scan can be converted into changes in height, providing a "topographic" image of the sample.

In an alternative imaging mode (*constant height imaging*) the feedback circuit is not used, but instead the tip is scanned by the piezo-crystal across the sample at a nominally constant height and variations in surface topography are recorded as changes in the tunneling current. This has the advantage of allowing faster imaging — which can be very useful in the observation of dynamic processes on surfaces. However, since there is no control over the height of the tip, there is a significant risk that it will crash into the sample if the surface height suddenly changes, and this may damage the tip making further imaging difficult or impossible. The difference between these two basic imaging modes (constant current and constant height imaging) are illustrated in Fig. 8.1.

Given the risk of damaging the tip by crashing it into a feature on the surface, it is fortunate that this key component of the STM is relatively simple to produce and replace. STM tips are typically made from W, Pt or Ir wire, and are formed either by mechanical cutting or by electrochemical etching. Since most STM studies are conducted under ultra-high vacuum (UHV) further tip formation may be performed in a vacuum environment, either by heating the tip, or by applying a large voltage pulse to it to cause field evaporation and (hopefully) tip sharpening. It is, however, probably fair to note that tip preparation remains something of a "black art".

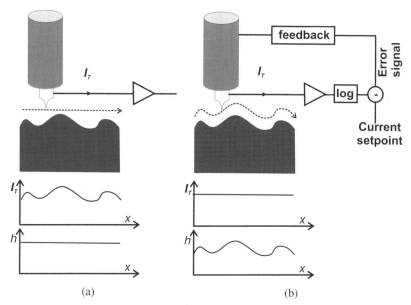

(a)   (b)

Fig. 8.1 Schematic diagrams of STM operation in (a) constant height mode and (b) constant current mode. It should be noted that whilst the graph of tunneling current, I_T, versus distance, x, in (b) shows a constant tunneling current, in reality the finite response time of the feedback circuit can lead to deviations of the tunneling current from the setpoint value.

Most STM studies are performed under UHV in order to provide a clean environment in which detailed surface structures can be studied, free from contamination or atmospheric oxidation. Whilst STM in air is possible, atomic resolution imaging of semiconductor surfaces is only feasible in a small minority of cases since most semiconductors form a native oxide instantly on exposure to air, and they are also likely to be covered by a contaminant layer consisting mainly of water. When the sample is put into the UHV system, it is desirable to remove the native oxide in order to access the atomic structure of the underlying semiconductor. For silicon and germanium, this is typically achieved by heating the sample to a temperature where the oxide layer evaporates.[8] Preparation of GaAs surfaces is more difficult, and it is often necessary to cleave the material under UHV to expose a clean, oxide-free surface.[9] For GaN, it is possible to clean the surface by heating it under an ammonia atmosphere.[11] However, most atomic resolution imaging of GaN has to date been carried out on samples grown by molecular beam epitaxy

R. Oliver

(MBE) in the same vacuum system as the STM, so that the sample is not exposed to air at all before imaging.[12]

Achieving UHV in an STM system requires some care, since if mechanical pumps were used the vibrations they would cause would significantly impact the achievable resolution of the system. Generally, this problem is overcome by the use of ion getter pumps, which have no moving parts. However, even without pump vibrations, vibration isolation is a key issue in the design of STM instrumentation. In order to measure atomic surface corrugations which may have an amplitude of 0.1 Å, or less, it is desirable to maintain the tip-sample distance with an accuracy of better than 0.01 Å.[6] The success of modern STM vibration isolation systems in achieving such low noise levels is a triumph of mechanical design.

8.2.3. Surface Reconstructions

One of the key concerns in STM of semiconductors is to understand the *reconstruction* of the surface at the atomic level. If you consider the imaging of a specific surface of a single crystal (such as the (100) surface), you might expect it to look a lot like the (100) plane of that crystal in the bulk of the material. However, in forming a free surface from the bulk material many bonds are broken, and this is a high energy situation. Hence, the atoms on the surface move into new configurations to lower the overall surface energy. For some metals, this rearrangement is fairly subtle. However, for most semiconductors, and for some metals such as Au and Pt, the rearrangement of atoms at the surface is substantial, so that the two dimensional (2D) surface unit cell has significantly different dimensions from the 2D unit cell which would describe the bulk-like termination of the crystal. This new arrangement of atoms at the surface is known as a *surface reconstruction*.

As an example of a surface reconstruction, we will briefly discuss the image of the Si(111) surface shown in Fig. 8.2. The understanding of the reconstruction of the Si(111) surface which was provided by STM was one of the first major triumphs of the technique,[10] and the beautiful pattern of atoms observed in Fig. 8.2 remains a classic sight for STM researchers. If one were to consider a bulk-like termination of the Si(111)

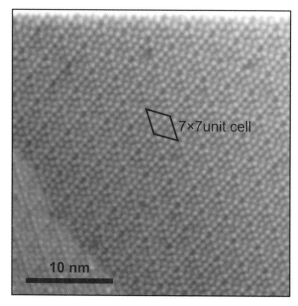

Fig. 8.2   STM image of a Si(111) 7 × 7 reconstruction. The 7 × 7 unit cell is indicated on the figure.

surface, one would expect the 2D unit cell to be a rhombus of side length 0.38 nm. However, the unit cell marked on Fig. 8.2 has the same rhombus-shape but has sides of length 2.7 nm, seven times greater than expected for the bulk-like termination. Hence, this structure is referred to as the Si(111) 7 × 7 reconstruction. Sometimes, the reconstructed unit cell has a different shape to that for the bulk termination unit cell. For example, for the Si(100) surface, the bulk periodicity would result in a square unit cell with a lattice parameter of 0.38 nm. The reconstructed surface has a rectangular unit cell, 0.76 nm by 0.38 nm in size. This is thus a 2 × 1 reconstruction, and it arises from the formation of surface dimers resulting from the pairing of dangling bonds present in the bulk termination. This dimerisation reduces the energy of the surface.[13]

The reconstructed unit cell may also have a different orientation to the unit cell of the bulk-like termination. One commonly occurring reconstruction of this type is the $(\sqrt{3} \times \sqrt{3})$ R30° structure, in which the reconstructed unit cell has the same shape as the bulk-termination unit cell, but is rotated by 30° relative to the bulk-termination unit cell and has lattice parameters a factor of $\sqrt{3}$ longer than the bulk-termination

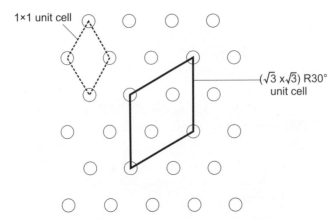

Fig. 8.3   Schematic illustration of the relationship of a $(\sqrt{3} \times \sqrt{3})$ R30° reconstruction to the underlying bulk-like termination of the crystal for a close-packed surface. The circles represent the positions of the atoms in the bulk termination, and the unit cell for this bulk termination is shown using dashed lines. The shape of the $(\sqrt{3} \times \sqrt{3})$R30° unit cell is superimposed on the figure in bold lines.

unit cell. Figure 8.3 shows a simple schematic of the (111) surface of a close packed crystal in its bulk termination, with the shape and size of a $(\sqrt{3} \times \sqrt{3})$ R30° unit cell superimposed.

Thus far the examples of reconstructions we have been given clean, elemental surfaces. For compound semiconductors, however, the observed reconstruction is strongly dependent on the stoichiometry of the surface. For example, Sec. 8.4 will provide a case study on research on the (0001) surface of GaN, where the surface reconstruction is seen to change from 2×2, to 5×5, to 6×4 and finally to 1×1 as the ratio of Ga to N on the surface increases.[14] Additionally, deposition of a small amount (a monolayer or sub-monolayer quantity) of another material onto a reconstructed surface can dramatically alter the reconstruction. One well-known example of this is the formation of a $(\sqrt{3} \times \sqrt{3})$ R30° reconstruction formed upon deposition of silver on Si(111).[15] The study of metal deposition on semiconductor surfaces is particularly interesting in the context of understanding the formation of contacts to semiconductor devices,[16] whilst the reconstructions formed in heteroepitaxial growth of one semiconductor upon another have practical relevance in the analysis of MBE growth, particularly for the growth of nanostructures such as quantum dots.[17]

8.2.4.   Cross-Sectional STM Imaging

Whilst SPM techniques intrinsically address the surfaces of materials, STM may also be used to study sub-surface nanostructures by cleaving a multilayer structure of interest *in vacuo* and then performing STM on the resulting cross-section. This approach is known as cross-sectional STM or XSTM, and has been extensively used to study sub-surface semiconductor nanostructures such as quantum wells or quantum dots. Such structures generally involve a layer of a lower bandgap material, sandwiched between layers of a higher bandgap material. The lower bandgap material often has a larger lattice parameter than the higher bandgap material and is thus under compressive strain. (Indeed, it is often this compressive strain which drives the formation of quantum dots.) Upon cleaving, some of this strain relaxes by the compressed material bulging outward, and hence there is a topographic change at the cleaved surface associated with the buried layer. STM images of the cleaved surface will thus be influenced by this topographic change and also by the differences in the surface LDOS between the two materials. For example, in imaging sub-surface InAs quantum dots in a GaAs matrix, images taken at negative sample biases (i.e. images of the As sublattice — see Sec. 8.2.1 above) will be influenced primarily by topography, whereas images taken at positive sample biases (i.e. images of the metallic sublattice) will be influenced by both topographic and chemical contrast.[18]

8.3.   ATOMIC FORCE MICROSCOPY

8.3.1.   Basic Principles of AFM

Atomic force microscopy (AFM) was invented a few years after STM by Binnig *et al.*,[19] and was intended to bring the power and resolution of STM to insulating materials through which it is impossible to flow a measurable tunneling current. Additionally, AFM was conceived from the start as a device capable of measuring not only the forces arising from interatomic interactions, but also other forces, such as those due to electric or magnetic fields.[19] AFM employs an ultrasharp tip mounted on a very small cantilever (typically a few hundred microns long and a few tens of microns wide), which is positioned very close to the sample surface. As in STM, the precision positioning of the tip is achieved using piezoelectric actuators.

The simplest way to operate the AFM, which is similar to the constant height mode in STM, would be to simply scan the cantilever over the surface, with the tip in contact with the surface and the base of the cantilever kept at a constant height, and then to measure the deflection of the cantilever as the tip moves up and down over topographic features. (The cantilever deflection is most commonly measured by monitoring the deviation of a laser beam reflected from the cantilever using a 4-sector photodetector.) However, as in STM, where in constant height mode the risk of damage to the tip is large, such an approach to AFM would provide no control over the forces on the tip and cantilever, making damage to both likely. Thus, although some very early measurements were taken in this way,[20] from the first,[19] it has been more usual to use a feedback circuit to control the height of the cantilever and to thus maintain a constant cantilever deflection, in much the same way that the feedback circuit controls the height of the STM tip in "constant current" STM. By controlling the cantilever deflection, the force on the cantilever is also maintained at a constant value, since for small deflections the two will be directly proportional to one another. As with STM, the feedback circuit adjusts the voltage applied to the piezo-scanner and calibration is again required so that these voltages may be accurately converted into changes in the height of the sample surface.

The mode of AFM operation described above, where the tip is in constant contact with the sample surface, and the feedback circuit is used to control the deflection of the tip, is usually called "contact mode". This is one of the two most commonly used AFM modes. The other is intermittent contact (IC) mode, often referred to as "tapping mode". IC mode uses a cantilever driven to vibrate at or close to its free space resonant frequency (ω_0). As the cantilever approaches the surface, the force gradient between the tip and the sample changes, resulting in a modification of the cantilever resonant frequency (ω) as follows:

$$\omega = \omega_0 \sqrt{1 - \frac{1}{k_0}\frac{dF}{dZ}} \tag{8.3}$$

where k_0 is the spring constant of the cantilever in free space. For a fixed drive frequency, this change in the resonant frequency, results in a change of the cantilever's amplitude of vibration, allowing the distance of the cantilever above the surface to be monitored based on measurement of this vibration amplitude. However, in typical imaging conditions in IC mode, this is not the most important effect. Since the tip attached to the vibrating cantilever makes contact with the surface of the sample under examination at the bottom of each oscillation, the presence of the surface limits the oscillation amplitude. This damping effect is the most significant cause of changes in the vibration amplitude with distance from the surface in IC mode. (Further details of the tip-sample interaction need not concern us here, but the reader is referred to Burnham et al.,[21] for more information.) As in contact mode, a feedback circuit is employed with the aim of maintaining a constant tip-sample interaction, but here, instead of adjusting the cantilever's height to maintain a constant cantilever deflection, the height adjustments maintain a constant amplitude of vibration. Once again, appropriate calibration data is needed to transform the voltages applied to the piezo-scanner into accurate information about the variations in sample topography.

IC mode has one major advantage over contact mode. In contact mode, whilst the normal force between the tip and the sample may be controlled and limited, the motion of the cantilever over the surface nonetheless gives rise to a lateral or friction force, which can result in damage to soft samples. In IC mode, the tip is not continuously in contact with the sample, but only briefly taps the surface at the bottom of each swing, resulting in greatly reduced frictional forces. Here, we are concerned with the imaging of semiconductors, which are generally fairly hard materials, and so damage to the sample in contact mode is not usually a major concern. However, frictional forces also tend to wear the tip, making it less sharp and thus (as we will see in the next section) reducing the resolution of the microscope. Hence, IC mode is very popular for topographic imaging in semiconductors. It should be noted that a further imaging mode is also possible — non-contact mode — in which the tip does not contact the surface, but vibrates just above it, and only the impact of local force gradients on the natural frequency can be used to assess the tip-sample interaction. However, most AFM imaging is performed in air, and many surfaces are thus coated in an absorbed layer of atmospheric contaminants — mostly water. In non-contact mode, the tip thus hovers above this layer, and that limits the resolution that can be achieved in imaging the true sample surface. In UHV, however, non-contact mode AFM may be used to achieve atomic resolution.[22]

AFM is often used to assess rougher samples than would commonly be imaged in STM, with feature heights over a micrometer being measured in some applications.[23] Such samples often include regions where the height changes very quickly (i.e. with very steep slopes) and in these circumstances, the finite response time of the feedback circuit often results in significant deviations of the tip-sample interaction from the setpoint (either the deflection setpoint in contact mode, or the amplitude setpoint in IC mode). This leads to errors in the height data. However, the errors in maintaining the setpoint are usually recorded, and can be displayed as an image which provides useful insights into the nature of the surface. Such images — known as deflection error images in contact mode, or amplitude error images in IC mode — highlight slopes on the surface, particularly in the fast scan direction, which the eye finds easy to interpret as relating to sample topography, and which may sometimes reveal details which whilst present in the topographic data set are difficult to spot visually. An example is given in Fig. 8.4 which shows a large GaN island grown on a sapphire substrate. In the amplitude error image (Fig. 8.4(b)) it is easier to pick out details of the large island's three-dimensional (3D) shape and the smaller structures which surround it than it is in the topography image (Fig. 8.4(a)). A similar effect could be achieved by taking a derivative of the topography image.

8.3.2. Topographic Resolution of AFM

AFM achieves excellent vertical resolution. The main factor limiting the resolution which can be achieved is noise — from electrical, acoustic and mechanical sources — but modern microscopes routinely achieve sub-Ångstrom noise levels. The lateral resolution of the AFM is

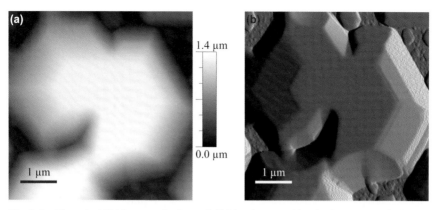

Fig. 8.4   IC mode AFM images of a large GaN island on sapphire. In the topography image (a) it is more difficult to see some of the details than in the amplitude error image (b).

determined by the size of the apex of the tip. The topography measured in AFM is sometimes (incorrectly) described as a convolution of the 3D shapes of the tip and the sample. Mathematically, in fact the image formation process in AFM should be described as a "dilation".[24] A simplified picture of the impact of the finite tip size on the measurement of typical features, such as 3D islands, in AFM is given in Fig. 8.5, in which the same AFM tip — depicted as a hemisphere — is shown interacting with two different small islands. The islands have the same width and sidewall angle but different heights, and the error in measurements of the width of the island (Δw) is seen to be dependent on the island height, and whether the tip contacts the island sidewall or the island corner.[25] This is illustrative of a general principal: the error in width measurements in AFM, depends not only on the size and shape of the tip used in measurement, but also on the size and shape of the object measured. Since, the shape of the object is usually unknown, and the tip shape may not be known precisely either, it is usually mathematically impossible to recover the true shape of the object from AFM data.[24]

Another important point, which is sometimes forgotten, is that despite the generally excellent vertical resolution of AFM, the finite tip size can also affect the AFM measurements in the vertical direction. As illustrated schematically in Fig. 8.6, in the imaging of a close-packed array of nanostructures, the tip may not be able to penetrate

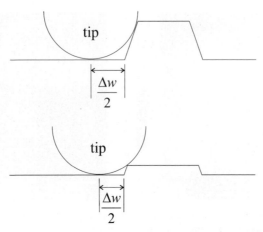

Fig. 8.5 Simple schematic showing the interaction of the same hemispherical AFM tip with two islands. The islands have the same width and sidewall angle but differ in height. This difference in height results in a difference in the error (Δw) in the measurement of their widths.

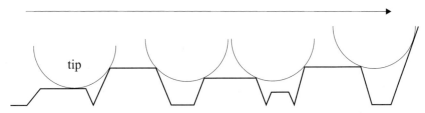

Fig. 8.6 Simple schematic of an AFM tip scanning over an array of densely packed nanostructures, showing that in several places the tip is unable to penetrate the gaps between the nanostructures.

into the bottoms of the gaps between the structures, resulting in an underestimate of the nanostructures' heights. Whilst the apex of a typical silicon AFM tip may be treated approximately as a hemisphere 5–10 nm in diameter, the overall shape of the tip on a broader scale is usually roughly pyramidal (as illustrated in Fig. 8.7). In the measurement of deep trenches (an important application of AFM in the characterization of semiconductor device processing) not only the apex of the tip, but also the slanting sidewalls of the pyramidal structure will interact with the sample, and this will impact the ability of the tip to penetrate into the bottom of a trench.

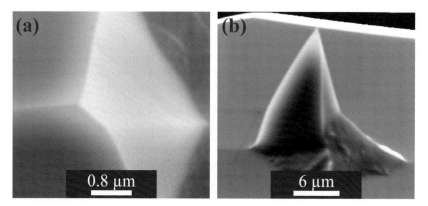

Fig. 8.7   Scanning electron microscopy images of a typical AFM tip: (a) from above and (b) from the side at an angle of 60° to (a).

8.3.3.   AFM-Based Techniques for the Electrical Characterization of Semiconductors at the Micro- and Nano-scale

This section will introduce some of the AFM-based techniques most commonly used in the electrical characterization of semiconductors. A more detailed review of this topic, which also describes a number of additional techniques, is given in Oliver.[25]

8.3.3.1.   Scanning capacitance microscopy

Scanning capacitance microscopy (SCM) is a contact mode technique, in which a metal-coated tip is scanned over an oxidized semiconductor surface.[26] The tip-sample system forms a nanoscale metal-oxide-semiconductor capacitor, the capacitance of which is monitored using an ultra-high-frequency capacitance sensor. An alternating bias is applied to the probe, resulting in alternating accumulation and depletion of carries in the semiconductor just below the tip. In the simplest mode — known as open loop SCM — the change in capacitance (ΔC) in response to a fixed change in applied voltage (ΔV) is measured. The magnitude of the observed signal depends on the dopant density in the semiconductor. To understand this, consider the application of a negative bias to the tip as it scans over an n-doped semiconductor surface: electrons will be repelled from the tip, creating a depletion region, and hence decreasing the measured capacitance (since the distance between the effective "plates" of the capacitor — the metal on the tip and the conductive semiconductor material — has increased). If the semiconductor is only lightly doped, the application of a negative bias will lead to the formation of a large depletion region, and the observed change in capacitance will be large. For the same applied bias, a smaller depletion region will form in a more heavily doped semiconductor and hence the change in capacitance will be smaller. In SCM an alternating bias is applied, so for an n-type semiconductor this depletion occurs in the negative half of the voltage cycle, whilst in a p-type semiconductor depletion occurs in the positive half of the cycle. Hence, the *phase* of the capacitance response to the applied voltage can be used to assess the carrier type, whilst the amplitude of the change in capacitance with applied voltage (dC/dV), depends on the carrier density.

In open-loop SCM, the applied alternating bias is fixed, and the dC/dV amplitude varies. Higher dC/dV amplitudes imply lower carrier densities in the semiconductor. In an alternative mode (closed-loop SCM), a fixed capacitance setpoint is chosen, and a feedback circuit is used to maintain that setpoint by varying the applied AC bias. The magnitude of the required AC bias is then used as a measure of the carrier density. A higher AC bias is required to deplete regions with higher carrier concentration sufficiently to achieve the desired capacitance change, so that in closed-loop imaging a high output corresponds to a region of high carrier density. Since in semiconductors the number of carriers in the material is related to the number of dopants present, SCM is often referred to as a dopant contrast imaging technique.

8.3.3.2. Scanning spreading resistance microscopy

Like SCM, scanning spreading resistance microscopy (SSRM) is a contact mode technique, often used to assess dopant densities in semiconductors. It is one of a family of techniques in which a constant bias is applied to the tip, and the resulting tip-sample current flow is measured. (Other techniques in this family include conductive AFM (C-AFM) and tunneling AFM (TUNA). The main differences between these techniques are not in the physics of their operation but in the sensitivity range of the hardware used to monitor the current flow.[25] In TUNA, for which very high sensitivity amplifiers are used to measure very small current flows, the mechanism of current flow may involve tunneling through an oxide layer, but this is not always the case). Whilst in SCM, low contact forces may be used, and the tip is separated from direct contact with the semiconductor by an oxide layer, in SSRM higher contact forces are used in order to push the tip through the oxide layer and allow direct contact with the semiconductor. For this reason, it is common in SSRM to use tips coated with doped-diamond rather than with metal, since the diamond coating provides a very hard surface which is stable under the application of large vertical and lateral forces.

Considering the semiconductor sample as a semi-infinite uniformly-doped slab of resistivity ρ that makes an Ohmic contact with a non-penetrating circular probe of radius a, the measured resistance, R, will be dominated by the spreading resistance,[27] so that:

$$R = \frac{\rho}{4a} \qquad (8.4)$$

This simple model ignores the contact resistance between the probe and the sample, which may be substantial and will be dependent on the probe shape, the surface state concentration and energy distribution (which will in turn be dependent on the local doping) and the force acting on the probe. The contact resistance alters Eq. (8.4) as follows:

$$R = \frac{\rho}{4a} + R_{\text{barrier}}(\rho) \qquad (8.5)$$

$R_{\text{barrier}}(\rho)$, a function expressing the variation of the barrier resistance as a function of the semiconductor's resistivity, may be non-linear, resulting in a complex relationship between the measured resistance in SSRM and the semiconductor's resistivity. Hence, SSRM data is often compared to data from standard samples of known resistivity or known doping density as an aid to quantification. Similar calibration samples are also useful in SCM, as will be illustrated in Sec. 8.6.

8.3.3.3.   Kelvin probe force microscopy

Kelvin probe force microscopy (KPFM) differs from SCM and SSRM in that it is a technique based on IC mode AFM. As noted above, in IC mode the amplitude of the cantilever vibration is largely dominated by the damping caused by the tip tapping on the sample surface. In KPFM, the AFM is used to probe the more subtle changes in local force which arise when a biased tip scans over a surface across which the local potential varies. The impact of the tip on the surface and the resulting damping are undesirable. Hence, in KPFM each scan line is repeated twice. Firstly a normal IC mode scan line is recorded to ascertain the sample topography. Then, the cantilever is lifted a small distance away from the surface and scans out the profile of the sample topography recorded in the previous scan line in order to maintain a fixed tip-sample distance. Data recorded during this second scan line — often known as a lift scan — are thus approximately independent of the sample topography, and accurate electrical data may be taken.

On the lift scan line, a biased, metal-coated tip scans over the surface, a fixed distance from the surface and the tip and the surface do not come into contact. The tip and sample may be modeled very simply as a parallel plate capacitor. The energy, U, stored in such a capacitor is given by

$$U = \frac{1}{2}C(\Delta V)^2 \tag{8.6}$$

where C is the tip-sample capacitance and ΔV is the difference in potential between the tip and the sample. The resulting force on the tip, in terms of the tip-sample separation, z, is given by:

$$F = -\frac{dU}{dz} = -\frac{1}{2}\frac{dC}{dz}(\Delta V)^2 \tag{8.7}$$

In KPFM, no mechanical vibration is applied to the tip, but the bias applied to the tip has both a DC component and an AC component with frequency ω, close to the resonant frequency of the cantilever:

$$\Delta V = \Delta V_{DC} + V_{AC}\sin(\omega t) \tag{8.8}$$

Hence, the force on the tip is given by:

$$F = -\frac{1}{2}\frac{dC}{dz}\left(\Delta V_{DC}^2 + \frac{1}{2}V_{AC}^2\right) - \frac{dC}{dz}\Delta V_{DC} V_{AC}\sin(\omega t)$$

$$+ \frac{1}{4}\frac{dC}{dz}V_{AC}^2\cos(2\omega t) \tag{8.9}$$

In Eq. (8.9), there are three contributions to the force — a static contribution, independent of ω (F_{DC}), an oscillating contribution with frequency ω (F_ω) and an oscillating contribution with frequency 2ω ($F_{2\omega}$). F_{DC} results in a cantilever deflection, and can be difficult to detect.[28] The amplitude of the cantilever's vibration is more straightforward to monitor. Since the cantilever's vibration response at its resonant frequency will be much stronger than it's response at the second harmonic, to a first approximation, the cantilever amplitude of vibration, A, is directly proportional to F_ω. Hence:

$$A \propto \frac{dC}{dz} \Delta V_{DC} V_{AC}. \tag{8.10}$$

The application of an alternating bias will thus lead to vibration of the cantilever unless ΔV_{DC} is zero.

If the DC tip bias is adjusted to achieve zero cantilever oscillation, ΔV_{DC} will be zero implying that the local surface potential is equal to the DC bias applied to the tip. During imaging, a feedback circuit is used to continuously adjust the tip DC bias in order to maintain A at zero, and an image is built up using the required DC biases. Hence, the KPFM image is a quantitative map of the surface potential. (Of course, as with the use of feedback circuits in AFM topographic imaging, the feedback circuit may not be able to maintain the amplitude at exactly zero, and this leads to errors in the measurement.)

Sections 8.2 and 8.3 have outlined various key SPM techniques. The following section (4, 5 and 6) will provide examples of the application of some of these techniques in the form of case studies centring on GaN-based materials.

8.4. CASE STUDY 1: SCANNING TUNNELING MICROSCOPY OF GAN(0001) SURFACES

STM of GaN is challenging due to the limited conductivity of GaN thin films, particularly those grown on insulating substrates such as sapphire. Smith et al.[14] grew GaN for study by STM in an MBE system adjacent to their STM chamber, and performed transfer under vacuum to the STM chamber, avoiding the necessity of the careful surface preparation[11] which is needed when GaN exposed to air is studied in STM. Smith et al.[14] achieved sufficient conductivity by doping the GaN they grew with Si. A few monolayers of undoped GaN were grown on top of the doped material, and various reconstructions were then achieved by either annealing the as-grown surface under a nitrogen-containing atmosphere, or by depositing metallic gallium on the surface and then annealing. In this way the surface stoichiometry was adjusted, altering the equilibrium surface structure. The observed surface structures were complex and this case study will not endeavor to provide a complete account of the discoveries of Smith et al.,[14] nor of the work of other authors[29] who have studied similar surfaces. Instead the aim here is to illustrate the beauty of atomic resolution STM data, and the difficulties which are sometimes involved in its interpretation.

8.4.1. STM Observations of Reconstructions of the GaN(0001) Surface

Smith et al.,[14] observed various different reconstructions depending on the stoichiometry of the GaN surface. In order to measure the surface stoichiometry they used Auger Electron Spectroscopy (AES). (In AES, the sample is bombarded with electrons of energy 3–20 keV. This bombardment results in removal of a core electron — i.e. an electron from an inner shell — from atoms at or near the surface of the sample. The atom then relaxes via an electron from a higher level dropping into the core hole. The energy thus released results in the emission of a third electron — an Auger electron — the energy of which is detected. Different atoms produce different spectra of Auger electron energies, allowing AES to be used in estimating surface compositions). Figure 8.8 shows the relative stoichiometries of the various observed reconstructions ranging form a 2 × 2 reconstruction with a low Ga content, to a reconstruction denoted "1 × 1" with a high Ga content.

An STM image of a 2 × 2 reconstructed surface is shown in Fig. 8.9. The surface appears rather disordered, and only small regions clearly show an ordered 2 × 2 reconstruction. This makes assessing the stoichiometry of the 2 × 2 reconstructed regions difficult. AES suggests that the overall surface has a fairly low Ga content. However, given the

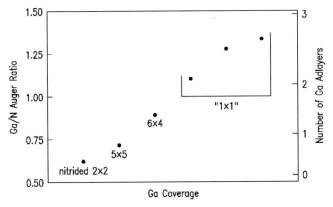

Fig. 8.8   AES data showing the relative stoichiometries of various reconstructions of the GaN(0001) surface. Reprinted from *Surface Science* Vol. 423, 70–84 A. R. Smith *et al.* Copyright (1999) with permission from Elsevier.

disordered nature of the surface, the stoichiometry might vary on a few nanometre scale, so that the small ordered region might have either a higher or a lower Ga content than the measured average stoichiometry. This makes understanding the detailed structure of the 2×2 reconstruction difficult, as will be further discussed in Sec. 8.4.2.

Figure 8.10 shows a more Ga-rich surface, which contains regions of the 5×5 reconstruction and regions of the 6×4 reconstruction.[14] The 6×4 is a well-ordered row-like structure and different domains, in which the rows are oriented along different $\langle 11\bar{2}0 \rangle$ directions, can be observed (labeled R1, R2 and R3). The 5×5 reconstruction seems less ordered, and it is more difficult to pick out different domains. Detailed analysis of the 5×5 structure, using variable voltages, suggests the existence of a well-defined 5×5 unit cell. However, the largest single 5×5 domain which has been observed is only a few unit cells in size.[14] The left hand side of Fig. 8.10 shows a region of 6×4 reconstruction and a region of 5×5 reconstruction within a single terrace. Interestingly, the average height of the 6×4 reconstruction is slightly lower than of the 5×5 reconstruction despite the fact that according to the AES data the 6×4 reconstruction is more Ga rich and the additional Ga atoms might be expected to increase the height of the surface. This difference in height is observed both when the sample is negatively biased (occupied states imaging, as shown in Fig. 8.10) and when the sample is positively biased

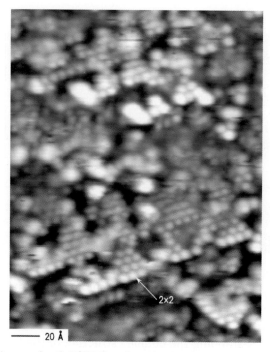

Fig. 8.9   STM image of a GaN(0001) surface showing small ordered areas of the 2 × 2 reconstructions. Reprinted from *Surface Science* Vol. 423, 70–84 A. R. Smith *et al.* Copyright (1999) with permission from Elsevier.

(unoccupied states imaging).[14] This makes it unlikely that the observed height difference is purely electronic in nature (i.e. purely an effect of the LDOS, not of the actual position of the atoms on the surface). Smith *et al.*,[14] attempt to explain the observed difference by suggesting a structural model for the 5 × 5 unit cell which is consistent with the data and which has a relatively open structure containing surface vacancies. They suggest that the 6 × 4 reconstruction may have a more densely-packed structure. If this were the case, then the finite size of the STM tip might make it difficult for the tip to penetrate into the vacancies on the 5 × 5 surface, and hence the surface might on average appear higher than its true height, since the dips in the surface might not be properly imaged, and the protuberances might be emphasized.[14] This example highlights the fact that whilst it is normally in AFM that the finite size of the tip is considered a significant concern, it does, of course,

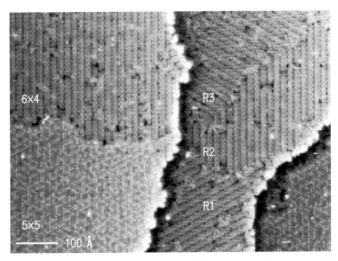

Fig. 8.10 STM image of a GaN(0001) surface showing areas of the 5 × 5 and 6 × 4 reconstruction. Reprinted from *Surface Science* Vol. 423, 70–84 A. R. Smith *et al.* Copyright (1999) with permission from Elsevier.

have a notable impact on STM imaging as well, despite the exponential dependence of tunneling current on tip-sample distance.

The last reconstruction observed by Smith *et al.*,[14] the "1 × 1", is particularly unusual. Whilst a 1×1 surface periodicity is observed in STM for this very Ga-rich structure, the protuberances observed in STM, are believed to relate to the underlying GaN lattice and not directly to the Ga layer which rests on the surface. This Ga layer is believed to consist of approximately two atomic layers of metallic Ga which are in a fluid-like state at room temperature.[14] The existence of such a floating metallic layer in MBE growth of nitrides may be quite important in maintaining a smooth surface morphology.[30]

8.4.2. Insights into GaN(0001) Surface Reconstructions from Atomistic Modeling

As we have seen above, difficulties in determining the surface stoichiometry, and the finite size of the tip, can make detailed interpretation of STM data difficult. Various tools exist which aid in the interpretation of STM data, including the application of other complementary surface analysis tools, such as X-ray photoelectron spectroscopy (XPS).[31] One increasingly common tactic in the interpretation of STM data is the combination of STM with atomistic modeling. The general approach is to first look at the data and generate a plausible model of the surface. The electronic structure of such a model can then be solved using one of a number of methods such as the Hartree-Fock approximation, density functional theory (DFT), tight-binding methods or quantum Monte Carlo simulations. The STM image which would be expected for the suggested model may then be computed (possibly at several biases) and compared with the original STM image to see whether the model's electronic structure is consistent with the actual experimental results.[32] Techniques such as DFT also allow us to compute the overall energy of the model structure. Hence, if several models are proposed, one might expect the structure with the lowest energy to form. If this minimum energy structure also yields a simulated STM image consistent with experimental data, then this would provide some confidence that the model is a good approximation to the real surface structure.

Smith et al.,[14] used DFT to examine possible structures for the GaN 2×2 reconstruction. They initially considered an N-rich structure containing one N adatom and one Ga rest-atom. (Adatoms are atoms which are adsorbed on top of the last essentially complete layer of the crystal at the surface, whilst rest-atoms are atoms within that essentially complete layer. However, a rest-atom does not have an adatom sat on top of it, so both adatoms and rest-atoms have a dangling bond associated with them.) They investigated whether the energy of the structure would be lower with the N adatom on the H3 site or the T4 site. (The locations of the H3 and T4 sites are illustrated schematically in Fig. 8.11). The model using the H3 site (shown schematically in Fig. 8.12(a)), was found to have a significantly lower energy than the model using the T4 site.[14] An alternative structure with the same stoichiometry as the model shown in Fig. 8.12(a) is shown schematically in Fig. 8.12(b). This structure has been formed by removing one N and one Ga adatom from each 2×2 unit cell, leaving behind a Ga vacancy. This structure has a rather similar energy to the N adatom model, but Smith et al.,[14] suggested that

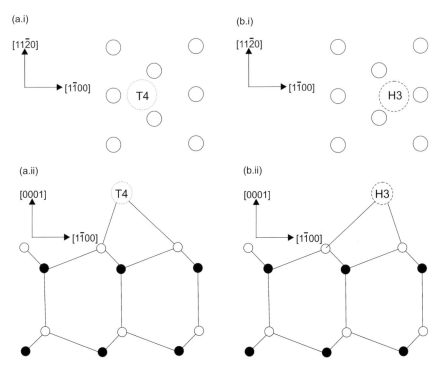

Fig. 8.11 Schematic illustration of the positions of the H3 and T4 sites. Ga atoms are open circles, whilst N atoms are closed circles. Dashed and dotted lines indicate the H3 and T4 sites respectively, which could be empty, or filled with either Ga or N atoms depending on the surface stoichiometry. (a) Position of the T4 site shown in (i) plan view, and (ii) cross section. (b) Position of the H3 site shown in (i) plan view and (ii) cross section. Cross sectional views are along the [11$\bar{2}$0] direction.

it was less consistent with their voltage-dependent imaging series for the real 2 × 2 reconstruction. Another alternative model was developed for surface with a more Ga-rich stoichiometry. The authors suggested that this model was less likely than the N adatom model, since the 2 × 2 reconstruction they observe is formed under N-rich conditions. However, they acknowledge that the difficulties in determining the local stoichiometry of the 2 × 2 structure make it difficult to be sure which model is most experimentally relevant.

This case study illustrates the power of STM in investigating the local atomic structure of semiconductor surfaces, but also indicates potential the drawbacks of the technique. We have seen that

R. Oliver

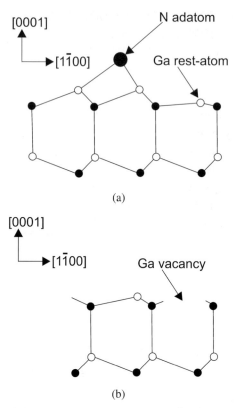

(a)

(b)

Fig. 8.12   Schematic cross sections viewed along the [11$\bar{2}$0] direction of the two possible low energy structures which Smith *et al.*,[14] suggested for the 2 × 2 reconstruction of the GaN(0001) surface. Ga atoms are open circles, whilst N atoms are closed circles: (a) Reconstruction with N atom on an H3 site. (b) Reconstruction with a Ga vacancy. The two reconstructions have the same stoichiometry and similar energies, but (a) is more consistent with experimental voltage-dependent imaging data. More detailed versions of these schematics can be found in Ref. 14.

interpretation of STM images must be affected by both topographic and electronic considerations and by the finite size of the STM tip, and that surface stoichiometry — which is key to interpreting surface structure in compound semiconductors — is difficult to determine directly from STM, and must usually be investigated using other complementary techniques. Whilst atomistic modeling, provides a powerful route to understanding STM data, it cannot substitute for experimental determination of local stoichiometry.

8.5. CASE STUDY 2: TOPOGRAPHIC ATOMIC FORCE MICROSCOPY STUDIES OF NANOSCALE NETWORK STRUCTURES IN INGAN

As was mentioned in Sec. 8.1, GaN-based LEDs emit brilliant light, despite containing a very high density of dislocations. The active region of such LEDs consists of one or more InGaN quantum wells (QWs), and inhomogeneities in these QWs could lead to local potential fluctuations which would prevent carrier diffusion to dislocation cores. The structure of such QWs at the nanoscale has been hotly debated[33] and many different techniques have been used to investigate it. This section will outline one example in which AFM has proved useful in understanding QW structure, particularly when combined with other complementary techniques.

8.5.1. Sample Description

TEM examination of some commercial green LEDs has revealed that the QWs found in their active regions are discontinuous, with gaps in the InGaN QW filled by GaN.[34] The separation between the observed gaps is about 50–100 nm. Such gaps occur in MOVPE-grown green-emitting QWs when, following the growth of a thin InGaN epilayer, the sample is either annealed at the InGaN growth temperature (typically c$_9$.710°C) in an N_2 and NH_3 atmosphere, or subjected to a temperature ramp to a higher temperature, prior to the growth of the GaN capping layer which completes the QW structure. Whilst basic AFM cannot be used to study subsurface QWs, it can be used to study epilayers which are subjected to the same thermal treatments as the layers in the QWs, but which are quenched (i.e. cooled to room temperature quickly) at the end of the anneal or temperature ramp, rather than being capped with GaN. In this case study we will describe the analysis of samples subjected to the annealing process described above and samples subjected to a temperature ramp from 700 to 860°C. Finally, in Sec. 8.5.3, we will also discuss the impact of adding a small flux of H_2 to the atmosphere during sample annealing.

8.5.2.   Complementary AFM, TEM and APT Analysis of Nanoscale Network Structures in InGaN

Figure 8.13 shows the impact of the thermal treatments on a 3.6 nm InGaN layer. The sample in Fig. 8.13(a) was quenched immediately after growth, and exhibits closely spaced terraces with raised decoration at their edges.[35] A sample annealed for 480 s in NH_3 and N_2 at growth temperature (Fig. 8.13(b)) shows an interlinked network of InGaN strips separated by narrow troughs. A sample subjected to a temperature ramp from its growth temperature to 860°C shows a similar network structure, but the troughs between the InGaN strips are broader (Fig. 8.13(c)). The data suggest that the gaps seen in QWs analogous to these epilayers relate to the troughs between the InGaN strips seen in AFM. Based on the AFM data, one might expect that carriers in a QW would be trapped within the InGaN, but would be able to move around fairly freely within the network structure.

Hence, if the dislocations are randomly distributed in the QW, it is difficult to see how the network structure would significantly impede carrier diffusion to dislocations.

AFM can be used to examine dislocations in GaN, since when a dislocation line terminates at a surface, the dislocation line tension results in a small pit at the surface.[36] The size of the dislocation pits may be increased using chemical treatments.[1] However, this method is usually

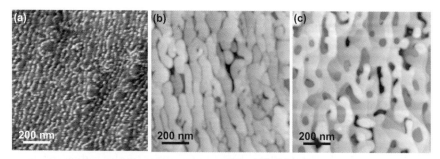

Fig. 8.13   AFM images of InGaN epilayers subjected to various treatments: (a) InGaN epilayer quenched immediately after growth. (b) InGaN epilayer annealed in NH_3 and N_2 at its growth temperature (710°C) for 480 s prior to quenching. (c) InGaN epilayer subjected to a temperature ramp under NH_3 and N_2 from its growth temperature to 860°C prior to quenching. Adapted from Refs. 34 and 35.

applied to flat, terraced GaN surfaces. On the rougher surface of an InGaN network structure, with many troughs, it becomes difficult, if not impossible, to identify the shallow pits associated with dislocations. Hence, in order to identify the relationship between the InGaN network structure and the location of dislocations, an alternative technique — bright-field TEM — was applied. Datta et al.[37] developed a bright-field multi-beam plan-view TEM imaging technique in which the beam is oriented along a $\{1\bar{2}1\bar{3}\}$ zone axis in order to reveal all types of dislocation in GaN. In such images, the InGaN network structure is also visible, and by taking images down different $\{1\bar{2}1\bar{3}\}$ zone axes, it is possible to identify which ends of the observed dislocation lines intersect the as-grown surface of the sample. Hence, it was shown that $90 \pm 10\%$ of the dislocations intersect the network structure either between the InGaN strips or at their edges, where the InGaN layer is thinner.[34] Following overgrowth of an InGaN epilayer with GaN to form a quantum well, this structure would thus provide a potential barrier to prevent excitons trapped in the InGaN network diffusing to dislocations cores.

Although AFM images provided the initial impetus for detailed characterization of the network structure, they were not able to provide the required information about the distribution of defects, and also could not provide access to compositional variations within the InGaN strips. Such compositional variations were initially examined using energy dispersive X-ray (EDX) analysis in the scanning transmission electron microscope (STEM). These studies indicated that the InGaN strips were richer in indium at their centres than at their edges.[34] More recently, three-dimensional atom probe tomography (APT) has been used to confirm these findings.[38] In APT, a needle-shaped sample with an apex radius of less than 100 nm is held at high voltage, so that the electric field at the apex is almost sufficient to ionize and field-evaporate atoms from the surface. In the analysis of semiconductors, field-evaporation is triggered by pulses of thermal energy provided by a laser. Evaporated ions are projected by the radial electric field on to a position sensitive detector and their chemical identities are determined by time-of-flight mass spectrometry. A three-dimensional atom map is then reconstructed from the data.[39] Whilst the first APT data sets recorded on GaN-based materials had a field-of-view whose lateral extent (about 20 nm)[40] was rather small in comparison to

the spacings of the InGaN strips in network structures, more recent measurements have been performed using a local electrode atom probe (LEAP)
with a larger field of view, allowing InGaN strips in green-emitting multiple QW structures and the gaps between them to be identified. The APT
data confirm the STEM-EDX data which suggest the InGaN strips are more
indium-rich at their centres than at their edges.[38]

8.5.3.   More Advanced Analysis of AFM Data from Network
Structures: Power Spectral Density Functions

Sometimes the "analysis" of AFM images involves nothing more complex than looking at a picture. For example, the images in Fig. 8.13
suggest that QWs with gaps consist of a network of interlinking strips,
whereas TEM cross-sectional images showing QWs with gross fluctuations in width had previously been interpreted as involving isolated
islands or disks of material.[41] Hence, the AFM image, in and of itself,
was enough to challenge a scientific hypothesis in the literature. However, more detailed quantitative analysis is clearly possible. Considering
the images in Fig. 8.13, one could obviously measure the widths of
the strips of InGaN, the depths of the troughs or another dimension of
interest. (In measuring the depths of the narrow troughs observed in
Fig. 8.13(b), it would be necessary to consider whether the tip could
penetrate to the bottoms of the troughs, and hence whether any measurement performed would be accurate.)

Rather than measuring the depths or heights of individual features, a
more global measure of the variations in height on the surface might be
useful. One possibility might be just to take the mean of all the measured
heights in the AFM image. However, if we consider the height of the
surface as a mathematical function $S(\mathbf{r})$ where S is the measured surface
height at position vector $\mathbf{r}$, then the values of S which are recorded will
depend on the height of the surface above some arbitrarily defined zero.
Hence, most AFM data is recorded as $H(\mathbf{r}) = S(\mathbf{r}) - \langle S(\mathbf{r}) \rangle$ where $\langle S(\mathbf{r}) \rangle$
denotes the mean of $S(\mathbf{r})$, giving a data set for which the mean is zero.
Hence, the basic arithmetic mean is not an appropriate metric for the
global variation in surface height. Various functions are used to fulfil this
role, most commonly the root mean square (rms) roughness, R_q. If $H(\mathbf{r})$

consists of a line profile of length L, then R_q is defined as:

$$R_q = \left[\frac{1}{L} \int_0^L H^2(x)dx\right]^{1/2} \tag{8.11}$$

(The vector **r** would in generally have an x and a y component. However, for a line profile we consider only variations in height in one spatial direction, denoted x.) Real AFM data is not recorded as a continuous function, but as a series of discrete numbers at various points, so Eq. (8.11) is modified as follows:

$$R_q = \left[\frac{1}{N} \sum_{i=1}^N H_i^2\right]^{1/2} \tag{8.12}$$

where N is the number of data points recorded along the line. Alternatively, for an AFM image, consisting of an array of N_x by N_y height values, we would write:

$$R_q = \left[\frac{1}{N_x N_y} \sum_{i=1}^{N_x} \sum_{j=1}^{N_y} H_{ij}^2\right]^{1/2} \tag{8.13}$$

A number of points should be born in mind in the application of R_q. Firstly, height data can be affected by the finite size of the tip, as illustrated in Fig. 8.6, and this will lead to errors in the value of R_q. It should be noted that the measured value of R_q may be an overestimate or an underestimate of the true surface roughness, depending on the shape of the features on the surface.[42] Secondly, the measured value of R_q will depend on the size of the image being analyzed (so that images of the same surface measured at different image sizes may give different R_q values) and on the data processing routines that have been applied to the image. Hence, to use R_q to make comparisons between the roughness of a series of samples, one should compare images of the same size which have been processed in the same way. Lastly, one should note that whilst it may be possible to use the variations in the measured value of R_q with image size to gain information on variations in the lateral feature sizes on a measured surface (i.e. the variations in the x and y directions), a single value of R_q does not tell you anything about these lateral length scales. Indeed, for the function given in Eq. (8.11), two sine

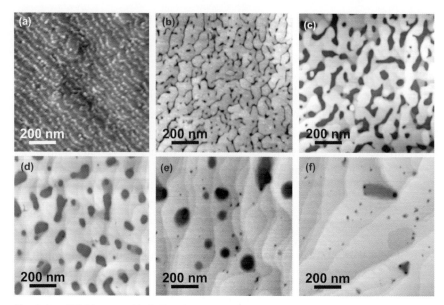

Fig. 8.14   AFM images of InGaN epilayers which have been annealed in NH_3 and N_2 with a small additional flux of H_2: (a) anneal time $t = 0\,s$, (b) $t = 60\,s$, (c) $t = 240\,s$, (d) $t = 360\,s$, (e) $t = 480\,s$, (f) $t = 960\,s$. Adapted from Ref. 43.

waves of the same amplitude but different frequency would give the same value of R_q.

With these limitations in mind, let us turn our attention to the data set shown in Fig. 8.14. Figure 8.14 shows various thin InGaN layers grown on GaN, which have been annealed for various lengths of time (from 0 s to 960 s). The anneal procedure applied was similar to that used for the sample in Fig. 8.13(b), but whilst the sample in Fig. 8.13(b) was annealed under a flow of N_2 and NH_3, the annealing atmosphere for the samples in Fig. 8.14 included an additional small flux of H_2, which alters the impact of annealing on the surface morphology.[43] For example, both Figs. 8.13(b) and 8.14(e) show samples annealed for 480 s, but whilst Fig. 8.13(b) shows very narrow troughs, Fig. 8.14(e) shows fewer, more rounded troughs. The evolution of the rms roughness with annealing time for the samples in Fig. 8.14 is shown in Fig. 8.15. The roughness measurements shown here are for images $3\,\mu m \times 3\,\mu m$ in size, which have all been processed in the same way. Four such images were taken from each sample, and mean rms roughnesses were found. The error

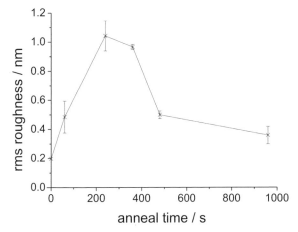

Fig. 8.15   The variation in the rms roughness of the InGaN surfaces shown in Fig. 8.14 with anneal time, t. Reprinted with permission from Ref. 43. Copyright 2009, American Institute of Physics.

bars in Fig. 8.15 represents the standard error on these means (i.e. the standard deviation of the recorded measurements divided by the square root of the number of measurements). It should be noted that these error bars do not include the impact of the finite tip size on the rms roughness measurement, and that this may be particularly profound for the samples which was annealed for 60 s (Fig. 8.14(b)), in which very narrow troughs are observed into which the tip may not be able to penetrate.

As the annealing time increases, Fig. 8.15 shows that the surface initially becomes rougher, but then becomes smoother again. In the remainder of this case study, this data set will be used to illustrate a more sophisticated analysis methodology which uses AFM data to access the mechanisms which smooth or roughen surfaces during deposition of material on a surface, or decomposition of material from a surface. For the samples in Fig. 8.14, high resolution TEM analysis[35] of the layer thicknesses in annealed and unannealed samples has shown that material is lost from the film during annealing — i.e. that the InGaN thin film is decomposing.

The method which will be illustrated here was initially developed by Tong and Williams,[44] who also review a number of other methods for the mathematical analysis of SPM images. If we again consider the height as a function of a position, but now also consider the different anneal times, t, we can treat each AFM data set as a function $H(\mathbf{r}, t)$, which as before will have a mean of zero. Tong and Williams[44] showed that it is possible to write an equation of motion for the growing (or decomposing) surface — essentially a kinetic rate equation — in terms of $h(|\mathbf{q}|, t)$, the radial average of the Fourier transform of $H(\mathbf{r}, t)$, where $|\mathbf{q}| = 1/|\mathbf{r}|$:

$$\frac{\partial h(|\mathbf{q}|, t)}{\partial t} \propto -c_n |\mathbf{q}|^n h(|\mathbf{q}|, t) + \eta(|\mathbf{q}|, t) \qquad (8.14)$$

In Eq. (8.14), n is a *smoothing exponent* which depends on the mechanism by which smoothing occurs, and c_n is a constant related to that smoothing mechanism. The function $\eta(|\mathbf{q}|, t)$ represents the random arrival of atoms at the surface (or for etching/decomposition the random removal of atoms from the surface) as a stochastic noise term. Tong and Williams[44] used Herring's[45] much earlier analysis of the mechanisms involved in sintering to suggest 4 possible smoothing exponents and related mechanisms: $n = 1$ for plastic flow, $n = 2$ for evaporation and recondensation, $n = 3$ for bulk diffusion and $n = 4$ for surface diffusion.

A solution to Eq. (8.1) may be found in terms of the radially averaged power spectral density (psd) of $H(\mathbf{r}, t)$, $g(|\mathbf{q}|, t)$:

$$g(|\mathbf{q}|, t) = \langle |h(|\mathbf{q}|, t)|^2 \rangle = \Omega \frac{1 - \exp(-2c_n |\mathbf{q}|^n t)}{c_n |\mathbf{q}|^n} \qquad (8.15)$$

A schematic of the form of this function is shown in Fig. 8.16 as a log-log plot. This illustrates how the $\mathbf{q}$-dependence of the spectral density depends on the smoothening mechanism.

If the only mechanism by which the surface roughens is by the random arrival of atoms at the surface during growth (or the random loss of atoms from the surface during decomposition), Eq. (8.14) will suffice. However, thin films often grow via the formation of islands — a non-stochastic roughening mechanism. Tong and Williams[44] suggested that island growth could be included in their analysis by considering either the roughening of the surface by deposition of material onto existing islands (with an exponent, n, of 1), or roughening of the surface by random

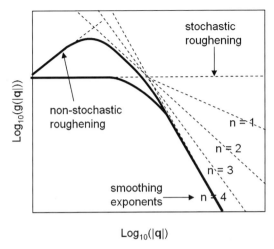

Fig. 8.16   Schematic diagram showing the influence of roughening and smoothening mechanisms on the psd function $g(q)$. Reprinted with permission from Ref. 43. Copyright 2009, American Institute of Physics.

arrival of atoms which then diffuse to existing islands (with an exponent, n, of 3). (Alternatively, for decomposition of material, one would consider, for example, roughening by *loss* of material from existing *pits*, with an exponent, n of 1, etc.). However, the signs of the constants c_n are different for roughening and smoothing. Hence, Tong and Williams rewrote Eq. (8.14) as follows:

$$g(|\mathbf{q}|, t) = \langle |h(|\mathbf{q}|, t)|^2 \rangle = \Omega \frac{\exp(2\sum \chi_n |\mathbf{q}|^n t) - 1}{\sum \chi_n |\mathbf{q}|^n t} \qquad (8.16)$$

The summation of terms in $\chi_n |\mathbf{q}|^n t$ indicates that a variety of roughening and smoothening mechanisms may operate simultaneously. The constants χ_n will be positive if the exponent, n, relates to a net roughening mechanism and negative where n relates to a net smoothening mechanism. For $n = 1$ and $n = 3$, there exist *both* a related smoothing mechanism and a related roughening mechanism. In these cases, a positive value of χ_3 (for example) would imply that there *is* roughening via the diffusion of material onto pre-existing islands, but would *not* necessarily imply the *absence* of smoothening by bulk diffusion. However, the positive value of χ_3 would imply that the roughening mechanism has more effect on the surface morphology than the smoothening mechanism. For $n = 1$ and $n = 3$, the impact of non-stochastic roughening may thus mask the impact of the relevant smoothing mechanisms — or vice versa. The impact of the non-stochastic roughening mechanisms on the form of $g(|\mathbf{q}|, t)$ is shown in Fig. 8.16.

This formalism may be applied in two ways to analyze real psd functions calculated from AFM data sets. One simple approach is to calculate the dominant smoothening and roughening exponents by fitting a straight line to the linear portions of the log-log plot of $g(|\mathbf{q}|, t)$. The modulus of the gradient of this line for large $\mathbf{q}$ — i.e. small length scales — then gives the approximate smoothing exponent whilst the gradient of the line at small $\mathbf{q}$ gives the roughening exponent. Alternatively, Eq. (8.15) may be fitted to the psd data to find values for the constants, χ_n.

For the samples shown in Fig. 8.14, the psd functions have been calculated using data from four $3\,\mu m \times 3\,\mu m$ images of the samples annealed for between 60 and 480 s (Fig. 8.17). For all the calculated psd functions, the gradient of the straight line portion of the log-log plot at high $\mathbf{q}$ is between 3.5 and 4.5, suggesting that the dominant smoothing exponent is $n = 4$, which would suggest that smoothing occurs via surface diffusion.[43] For the samples annealed for 240 to 480 s (Figs. 8.17(b)–(d)), the small straight line portion at low $\mathbf{q}$ has a gradient of between 1.1 and 1.3, suggesting that roughening of the surface is non-stochastic and occurs primarily by loss of material from existing pits.[43] For the sample with the shortest anneal time (Fig. 8.17(a)), the straight line portion at low $\mathbf{q}$ has gradient which is close to zero (perhaps suggesting stochastic roughening) but which is actually slightly negative. It should again be noted that Fig. 8.14(b) shows this sample to have very narrow troughs, so over-interpretation of the resulting psd function should probably be avoided since the AFM data may not be a good expression of the true surface morphology if the tip does not penetrate the bottoms of these troughs.[43]

Oliver et al.,[43] also fitted the full function given in Eq. (8.16) to the psd functions shown in Fig. 8.14(b) and showed that a good fit could be achieved with R^2 values (indicating the goodness of fit) being in excess of 0.99. Additionally, given the simple analysis above

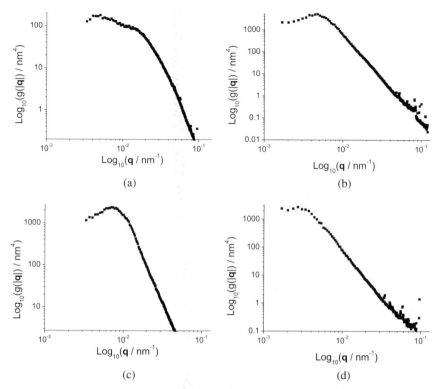

Fig. 8.17   psd functions calculated for $3 \times 3 \, \mu m^2$ AFM images of annealed InGaN epilayers: (a) anneal time $t = 60\,s$, (b) $t = 240\,s$, (c) $t = 360\,s$, and (d) $t = 480\,s$. Adapted from Ref. 43.

which implied that the dominant smoothing exponents for roughening and smoothening were 1 and 4 respectively, they applied a restricted model, in which the values of χ_2 and χ_3 were set to zero, so that the resulting function only depended on smoothening by surface diffusion and roughening by loss of material at pre-existing pits. Again, good fits were achieved, even with fewer parameters, with R^2 values in excess of 0.96. This provides some confidence in the assertion that these two mechanisms are key to the development of the observed surfaces during annealing.

Overall, this example illustrates the utility of mathematical analysis of AFM data in providing rather detailed information on nanoscale surface morphologies, not just in terms of what structures are present on the surface, but also in terms of how those structures formed. (For another example of application of sophisticated mathematical analysis to AFM data from GaN-based semiconductors, the reader is referred to Moram et al.,[2] in which spatial analysis techniques, originally developed for ecology, are used to assess the spatial distributions of dislocations at GaN surfaces, and to thus draw conclusions about origins of dislocations in these materials.) In addition to demonstrating the power of rigorous mathematical analysis, this case study has also shown that the combination of AFM data with other techniques — such as was discussed in Sec. 5.2 — can yield a still richer range of information.

8.6. CASE STUDY 3: SCANNING CAPACITANCE MICROSCOPY OF UNINTENTIONAL DOPING IN GAN

Another challenge in the development of GaN-based devices — particularly electronic devices such as HEMTs — is control of unintentional doping. Epitaxially-grown GaN is often found to be unintentionally n-doped due to the incorporation of impurities such as oxygen. However, doping does not occur uniformly throughout the epitaxial layer. Instead, a layer of material with a high density of unintentional dopants is often found close to the GaN/substrate interface. Additionally, as will be demonstrated, the thickness of this layer is often non-uniform on a length scale of microns or less. Hence, SPM-based electrical characterization allows access to details which cannot be discovered by more conventional macroscopic techniques such as capacitance-voltage (C-V profiling) or Hall probe measurements. This Case Study will illustrate the use of scanning capacitance microscopy to quantify unintentional doping in GaN.

8.6.1. Sample Description

The first challenge in developing the SCM technique for quantitative measurements in n-type GaN was to identify the range of dopant densities over which a capacitance response could be observed in SCM, and the relationship between the SCM dC/dV amplitude signal and the dopant density within that range. SCM has been rather extensively studied for application to Si,[46] but up until recently[47] GaN had not been systematically addressed. Given the profound differences in physical properties between GaN and Si (in terms of, for example, band gap, defect density and crystal structure) it was imperative to quantify the basic performance of the SCM technique using well-characterized structures, prior to attempts to explore unknown samples. In SCM, the dC/dV amplitude signal in open loop mode is expected to increase monotonically as the carrier density in the material increases, up to some detection limit. However, in studies on Si, this relationship is not always observed — sometimes the signal first increases and then decreases again with decreasing dopant density.[48] This "contrast reversal" is attributed to the effect of charge trapping in the oxide layer at the Si surface, and requires the application of a DC bias to offset the field due to the trapped charges and achieve interpretable data.[48] It was thus necessary to find out whether a similar effect occurs in GaN, and to find imaging conditions which allow interpretable data to be recorded.

The sample used to examine the relationship between dopant concentration and SCM dC/dV amplitude signal consisted of a GaN epitaxial layer grown by MOVPE on sapphire. The first ca. 5 μm of material were nominally undoped. On this GaN/sapphire "pseudosubstrate" a series of eight 200 nm thick doped layers were grown, separated by 200 nm thick undoped "spacers". The dopant density in the doped layers ranged from 5×10^{17} cm^{-3} to 1×10^{19} cm^{-3}, and the dopant densities were checked using secondary ion mass spectrometry (SIMS). The resulting SIMS data is shown in Fig. 8.18. For SCM studies, the sample was cleaved to allow images to be taken in cross-section. The details of the cleaving method used are described by Sumner *et al.*[49]

Once the relationship between SCM dC/dV amplitude and dopant density had been established (see Sec. 8.6.2 below), the SCM technique was used to quantify the unintentional doping in various structures, including an unintentionally-doped layer at the GaN/sapphire interface of the test structure described above (see Sec. 8.6.3) and unintentional doping in a more complex GaN/sapphire sample grown by epitaxial layer overgrowth (ELOG) (see Sec. 8.6.4).

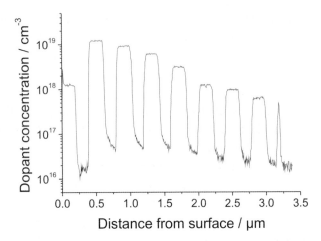

Fig. 8.18   SIMS dopant concentration data for a sample used to examine the relationship between dopant concentration and SCM dC/dV amplitude signal in GaN. Adapted from Ref. 47.

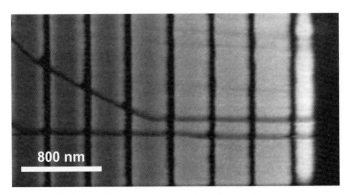

Fig. 8.19   SCM dC/dV amplitude image for the sample for which SIMS data is displayed in Fig. 8.18. Adapted from Ref. 47.

8.6.2.   SCM Data on a Test Sample

Figure 8.19 shows a typical SCM dC/dV amplitude image of the test sample. Seven differently-doped layers can be seen, along with a thin unintentionally-doped layer. This unintentional doping arises because after the first five microns of undoped GaN were grown, the sample was removed from the growth system for inspection, and then returned to the growth system for overgrowth of the doped multilayer. The interface between the GaN grown initially and the GaN grown after the wafer was removed from and returned to the MOVPE reactor is sometimes called the "regrowth interface". Unintentional doping at the regrowth interface has also been identified using other techniques.[50]

Figure 8.19 also shows that the doped stripes appear to be rather wider than the expected 200 nm. There are two reasons for this: firstly the finite width of the tip will tend to broaden electrical contrast features, just as it broadens topographic features. Secondly, SCM is sensitive to the presence of charge carriers in the material, rather than to the presence of the dopants themselves. Some carriers will diffuse out of the doped layers into the undoped spacers, and these carriers will affect the SCM signal. Since the carrier density is lower in the parts of the undoped region into which the carriers have diffused than in the doped stripe, the SCM dC/dV amplitude signal will be higher. Hence, some of the stripes in the image, particularly those with high doping densities, show bright regions at their edges. The impact of the finite size of the tip on the widths of features in SCM of GaN has been studied in more detail by Sumner *et al.*,[47] who showed that for platinum-iridium coated tips the error in measuring the widths of features is typically ca. 80 nm, and that it is possible to identify doped features with widths down to 10 nm.

The variation in SCM dC/dV signal with doping density is shown in Fig. 8.20. A monotonic decrease in the SCM signal with increasing

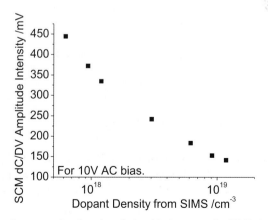

Fig. 8.20   Calibration curve showing the relationship between the SIMS data from Fig. 8.18 and the relevant SCM dC/dV amplitude data for SCM data recorded using a 10 V AC bias. Reprinted with permission from Ref. 47. Copyright 2008, American Institute of Physics.

dopant density is observed over the range of dopant densities studied here, with no apparent contrast reversal. Doped multilayers such as that in this test sample are thus suitable structures to provide calibration data for the quantification of dopant densities in unintentionally-doped features, provided the dopant density lies within the range studied.[47] This approach will be illustrated in the next section.

8.6.3. Quantification of SCM Data from Unintentionally-Doped Material

Figure 8.21 shows another SCM dC/dV amplitude image of the test structure studied in the preceding section. However, this image shows not only the doped multilayer but also the GaN underlying the multilayer and the GaN/sapphire interface. A bright layer of varying thickness is seen adjacent to the GaN/sapphire interface. By using the SCM data from the intentionally-doped multilayer to plot a calibration curve similar to that seen in Fig. 8.20, and then comparing the dC/dV amplitude signal from the unintentionally-doped interface layer to this calibration curve, it is possible to estimate the doping density in the unintentionally doped layer to be $1.7 \times 10^{18} \pm 1 \times 10^{17}$ cm^{-3}.[48] A sample similar to that show here but without the doped multilayer grown on top was also examined by SIMS, and an oxygen containing region close to the GaN/sapphire interface was identified, with an oxygen concentration of around 2.4×10^{18} cm^{-3}.[51] Given that oxygen is a shallow donor in GaN,

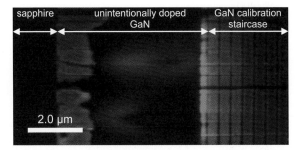

Fig. 8.21   SCM dC/dV amplitude image of a GaN epilayer grown on sapphire. The GaN below the calibration staircase structure is not intentionally doped (nid), but the presence of carriers is nonetheless detected in a region adjacent to the GaN/sapphire interface. (Reprinted from Ref. 25, with permission from IOP Publishing Ltd).

the similarity of the oxygen concentration found by SIMS and the dopant density estimated in SCM provides some confidence in the calibration procedure.

It should be noted that the quantification described above utilized a dopant calibration multilayer grown directly on top of a region of interest (ROI). This allowed the calibration structure and the ROI to be included in a single image. It was found that if the calibration structure was imaged first and then the tip was moved out of contact with the surface and then brought back into contact with the surface to allow the ROI to be imaged, significant errors arose in the quantification of the dopant content in the unintentionally-doped layer, due to changes to the tip. Hence, for accurate quantification of dopant densities in GaN-based materials it is recommended to image a calibration structure and the ROI simultaneously. In other materials that could be achieved by gluing together two structures — one containing the ROI and the other containing the calibration material, and then polishing the two to achieve a flat cross-section so that both could be included in one image. However, for GaN, polishing appears to alter the electrical characteristics of the cross-sectional surface and makes it difficult or impossible to collect SCM data.[49] Hence, it is preferable to grow the calibration structure on top of the ROI, so that they can be cleaved and then imaged simultaneously.

8.6.4.   Application of SCM in Understanding Facet-Dependent Dopant Incorporation

This section will further illustrate the utility of SCM in assessment of unintentional doping in GaN by addressing a sample grown by a method called epitaxial lateral overgrowth (ELOG) which is used to reduce the density of dislocations in the material. The structure of the ELOG sample studied here is illustrated in Fig. 8.22 and, and the ELOG process employed will be explained with reference to this figure. Initially, a GaN seed layer is grown on sapphire, which may contain a high density of dislocations.[52] (Dislocations are shown as thin black lines in Fig. 8.22.) The seed layer is then taken out of the MOVPE growth system, and using conventional lithographic techniques a pattern of stripes in an amorphous material such as silicon nitride is applied to the surface forming a mask.

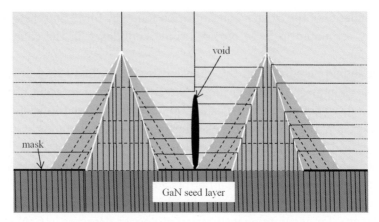

Fig. 8.22   Schematic structure of an ELOG sample studied by SPM. Fine black lines represent dislocations. Dashed lines represent the shapes of the stripes of GaN as it grows through the windows in the mask. Solid white lines mark the position of the edges of the stripes of GaN as the shape changes during low V:III ratio growth. Adapted from Ref. 52.

The sample is returned to the MOVPE reactor for overgrowth of more GaN. Deposition of material occurs in the windows between the masked regions. Dislocations can propagate from the seed layer in these window regions, but defects in the material under the mask are blocked from propagating. For the specific technique shown here, the growth conditions are selected such that the formation of inclined $\{11\bar{2}2\}$ side facets is favored. Hence, the GaN forms stripes of trapezoidal cross-section, bounded by $\{11\bar{2}2\}$ side facets and with a (0001) top facet (indicated by dotted black lines in Fig. 8.22). Where dislocations intersect the $\{11\bar{2}2\}$ side facets they tend to bend over, and propagate laterally. As the stripes grow, the $\{11\bar{2}2\}$ facets expand and the (0001) facet shrinks, until the stripe cross-section is triangular (indicated by white dotted lines in Fig. 8.22).[52]

At this point the growth conditions are changed, by altering the V:III ratio and introducing a flux of bis(cyclopentadienyl)magnesium (Cp_2Mg) in order to encourage lateral growth and re-formation of the (0001) facet. Hence the GaN from the window region overgrows the masks to form "wings" of low dislocation density. Where these wings coalesce a void is often formed, and laterally-propagating dislocations may terminate at this void. Alternatively, dislocations may meet each other and

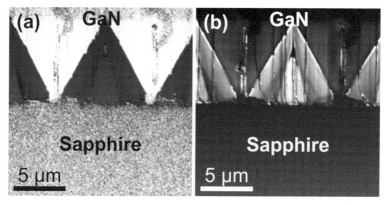

Fig. 8.23 Cross-sectional scanning capacitance microscopy data for an ELOG sample. (a) dC/dV phase data. (b) dC/dV amplitude data. Adapted from Ref. 52.

annihilate at the coalescence boundary, or may combine to form a vertically propagating dislocation. As is illustrated schematically in Fig. 8.22, the overall dislocation density is dramatically reduced with respect to the seed layer. The behavior of the dislocations shown schematically in Fig. 8.22 has been confirmed using weak-beam dark field TEM imaging in cross-section in the TEM.[52]

Figure 8.23 presents both SCM dC/dV phase data (Fig. 8.23(a)) and SCM dC/dV amplitude data (Fig. 8.23(b)) for the ELOG sample shown schematically in Fig. 8.22 examined in cross-section. SCM phase data gives information about the type of doping present in the sample, and for the imaging conditions applied here, p-type material is seen as white, n-type material as black and insulating material as grey and noisy (since it does not deplete under the applied bias and thus exhibits no fixed phase response).[49] Hence, Fig. 8.23(a) shows triangles of unintentionally n-doped material, corresponding to the triangles indicated by white dashed lines in Fig. 8.22. Between these n-type triangles the material is p-type. The presence of p-type doping is related to the use of Cp_2Mg to speed up the coalescence process;[53] Mg forms an acceptor state in GaN. However, the existence of measurable p-type doping is nonetheless surprising since in MOVPE growth, Mg usually forms an electrically-inactive complex with hydrogen and Mg-doped GaN must be annealed in a nitrogen atmosphere to activate the acceptors providing free holes. No such anneal step has been used here, but nonetheless p-type conductivity is observed. The presence of n-type conductivity is less surprising since either Si from the silicon nitride masking layer, or oxygen present as an impurity in the precursor or carrier gases might act as an n-type dopant.

Considering Fig. 8.23(b), the SCM dC/dV amplitude data, variations in the intensity of the amplitude signal are observed corresponding to variations in the dopant concentration. Where growth has occurred on the (0001) facet (corresponding to the area within the solid white triangle in Fig. 8.22), the image appears brighter, corresponding to a lower dopant concentration. Where growth has occurred on a $\{11\bar{2}2\}$ side facet, the SCM dC/dV amplitude signal is lower and also shows a reduction between the edge of the n-doped triangle and the area closer to the middle. This implies that dopant incorporation occurs more quickly on the $\{11\bar{2}2\}$ side facets than on the (0001) facets and that the first material to grow on the $\{11\bar{2}2\}$ side facets has a higher dopant concentration than subsequent material.[53] The variations in dopant concentration within the n-type region have been quantified by Sumner et al.,[53] who used a calibration structure grown on top of the ELOG material to aid quantification.

SCM here provides evidence of facet-dependent dopant incorporation within a single sample. This effect would be difficult to investigate using a macroscopic technique. (It is worth noting that whilst secondary electron imaging in the scanning electron microscope (SEM) is also sometimes used as a dopant contrast technique, attempts to use it to study this sample only showed contrast between the n- and p-doped regions. The differences in doping within the n-doped region were not visible in the SEM image.[47]) Studies of facet-dependent doping using SCM have also been performed in other materials systems — for instance to investigate dopant incorporation on different facets during the growth and fabrication of InP buried heterostructure lasers.[54] This case study and other similar studies illustrate the usefulness of AFM-based electrical characterization techniques in providing quantifiable information about dopant densities in two-dimensions on the micro- to nano-metre scale. In SSRM, efforts are even being made to extend the technique to three-dimensional nanoscale profiling for certain specific materials, particularly InP.[55]

8.7. CONCLUDING REMARKS

This chapter has introduced some key SPM techniques, and just a few of the applications of SPM to semiconductors. STM and AFM have provided scientists with an amazing vision of semiconductor materials at the nanoscale, and their invention could be considered the starting point of the nanoscience and nanotechnology revolution.[56] SPM techniques are enormously adaptable, and are now used to assess not only the topography of nanostructured samples, but a diverse range of other properties: mechanical, electrical and magnetic. However, as quickly as SPM has developed, semiconductor technology seems to evolve at an even more rapid pace, with the ongoing shrinkage of silicon-based electronic devices, the appearance of new semiconductors like GaN, and the development of a plethora of new techniques for the formation of nanoscale structures all placing increasing demands on materials characterization techniques.

In this context, it is unsurprising that this chapter has only been able to review a subset of the SPM techniques which are currently applied to semiconductors, and indeed new techniques are being developed all the time, some to assess very specific problems, whilst others will find wider application in the long term. In the application of any SPM technique however, the inexperienced scientist should bear some key points in mind: whilst the AFM in particular has developed to the point where systems are available which make data collection apparently extremely straightforward, such turn-key systems do not possess an inbuilt ability to optimize the imaging conditions for unusual or challenging samples, to spot artefacts or to interpret data. Hence, to record and understand high quality data, the user still needs a reasonable understanding of key components such as the feedback circuit, the piezo-scanner and the tip. The finite size of the tip, in particular, must be considered in all SPM measurements, as was noted in all the case studies in this chapter. With an increasing number of specialist tips being commercially-available, care should be taken to select the right tip for the job. As well as understanding the physics of SPM image formation, and the instrumentation used to achieve imaging, an effective microscopist must also understand the processing algorithms applied to their data — an area on which this chapter has barely touched. Meaningful analysis of AFM data requires appropriate data processing, but (as we saw in Sec. 8.5.3) can, when done properly, yield a surprising depth and breadth of information. SPM techniques can produce very "pretty pictures", but are perhaps at their best when our analysis goes beyond this to provide more quantitative assessment of a problem.

SPM techniques alone can yield some fascinating insights, but they are not without their limitations. Direct compositional analysis by SPM remains extremely difficult, and the assessment of sub-surface structures is only possible via cleaving to expose said structures or via specialist applications of techniques such as SCM.[57] In this context, the combination of SPM techniques with other methods for nanoscale characterization becomes increasingly vital. In Sec. 8.4, the use of compositional information from AES to complement STM data was described. In Sec. 8.5, both APT and TEM were used alongside AFM, and in Sec. 8.6, SIMS measurement of calibration samples was required to allow quantification of SPM data. Looking to the future, the combination of SPM with other techniques offers various opportunities. For example, electrical SPM techniques may be used to identify regions of device structures with poor electrical properties — such as current leakage spots in barrier materials.[58] Currently, identification of the structural defect leading to these leakage spots may be difficult, but with the increasing availability of focussed ion beam (FIB) microscopes, it should be possible to identify such a specific feature in SPM, use FIB to make a thin film cross-section at its exact location, and then use TEM to identify the structural defect.[25] Going beyond such approaches, the integration of SPMs into electron microscopes offers numerous exciting possibilities, although currently this strategy has only found niche applications.[59,60] Nonetheless, the uses to which scientists are putting SPM probes is ever-expanding, and appears to be only limited by our imaginations. SPM is already a key tool of nanoscience and nanotechnology, and has the potential, in the future, to take scientists to some unexpected places.

REFERENCES

1. R. A. Oliver, M. J. Kappers, J. Sumner, R. Datta, C. J. Humphreys, *J. Cryst. Growth.* **289**, 506 (2006).
2. M. A. Moram, R. A. Oliver, M. Kappers, C. J. Humphreys, *Adv. Mat.* **21**, 3941 (2009).
3. L. Solymar, D. Walsh, *Electrical Properties of Materials* 6th Edition, Oxford University Press, Oxford, pp. 40–42 (1998).
4. G. Binnig, H. Rohrer, C. Gerber, E. Weibel, *Phys. Rev. Lett.* **49**, 57 (1982).
5. J. Tersoff, D. R. Hamann, *Phys. Rev. Lett.* **50**, 1998 (1983).
6. J. A. Kubby, J. J. Boland, *Surf. Sci. Rep.* **26**, 61 (1996).
7. C. J. Chen, *Phys. Rev. Lett.* **65**, 448 (1990).
8. W. A. Hofer, A. S. Foster, A. L. Shluger, *Rev. Mod. Phys.* **75**, 1287 (2003).
9. R. M. Feenstra, J. A. Stroscio, J. Tersoff, A. P. Fein, *Phys. Rev. Lett.* **58**, 1192 (1987).
10. R. M. Hamers, R. M. Tromp, J. E. Demuth, *Phys. Rev. Lett.* **56**, 1972 (1986).
11. R. A. Oliver, C. Nörenberg, M. G. Martin-Fernandez, A. Crossley, M. R. Castell, G. A. D. Briggs, *Appl. Surf. Sci.* **214**, 1 (2003).
12. A. R. Smith, R. M. Feenstra, D. W. Greve, J. Neugebauer, J. E. Northrup, *Phys. Rev. Lett.* **79**, 3934 (1997).
13. R. M. Hamers, R. M. Tromp, J. E. Demuth, *Phys. Rev. Lett.* **34**, 5343 (1986).
14. A. R. Smith, R. M. Feenstra, D. W. Greve, M.-S. Shin, M. Skowronski, J. Neugebauer, J. E. Northrup, *Surf. Sci.* **423**, 70 (1999).
15. X. Tong, Y. Sugiura, T. Nagao, T. Takami, S. Takeda, S. Ino, S. Hasegawa, *Surf. Sci.* **408**, 148 (1996).
16. C. Norenberg, M. R. Castell, *Surf. Sci.* **601**, 4438 (2007).
17. I. Goldfarb, J. H. G. Owen, P. T. Hayden, D. R. Bowler, K. Miki, G. A. D. Briggs, *Surf. Sci.* **394**, 105 (1997).
18. O. Flebbe, H. Eisele, T. Kalka, F. Heinrichsdorff, A. Krost, D. Bimberg, M. Dähne-Prietsch, *J. Vac. Sci. Technol. B* **17**, 1639 (1999).
19. G. Binnig, C. F. Quate, C. Gerber, *Phys. Rev. Lett.* **56**, 930 (1986).
20. H. Heinzelmann, P. Grutter, E. Meyer, H. Hidber, L. Rosenthaler, M. Ringger, H. J. Guntherodt, *Surf. Sci.* **189**, 29 (1987).
21. N. A. Burnham, O. P. Behrend, F. Oulevey, G. Gremaud, P.-J. Gallo, D. Gourdon, A. J. Kulik, H. M. Pollock, G. A. D. Briggs, *Nanotechnology* **8**, 67 (1997).
22. R. Erlandsson, L. Olsson, P. Martensson, *Phys. Rev. B.* **54** R8309 (1996).
23. R. A. Oliver, M. J. Kappers, C. J. Humphreys, *Appl. Phys. Lett.* **89**, 011914 (2006).
24. S. Dongmo, M. Troyon, P. Vautrot, E. Delain, N. Bonnet, *J. Vac. Sci. Technol. B* **14**, 1552 (1996).
25. R. A. Oliver, *Reports on Progress in Physics* **71**, 076501 (2008).
26. C. C. Williams, *Annu. Rev. Mater. Sci.* **29**, 471 (1999).
27. P. De Wolf, T. Clarysse, W. Vandervorst, *J. Vac. Sci. Technol. B* **16**, 320 (1998).

28. P. Girard, *Nanotechnology* **12**, 485 (2001).
29. Q.-K. Xue, Q. Z. Xue, R. Z. Bakhtizin, Y. Hasegawa, I. S. T. Tsong, T. Sakurai, *Phys. Rev. Lett.* **82**, 3074 (1999).
30. B. Daudin, G. Feuillet, G. Mula, H. Mariette, J. L. Rouviere, N. Pelekanos, G. Fishman, C. Adelmann, J. Simon, *Phys. Stat. Sol. A* **176**, 621 (1999).
31. C. Tornevik, M. Gothelid, M. Hammar, U. O. Karlsson, N. G. Nilssong, S. A. Flodstrom, C. Wigren, M. Ostlin, *Surf. Sci.* **314**, 179 (1994).
32. G. A. D. Briggs, A. J. Fisher, *Surf. Sci. Rep.* **33**, 1 (1999).
33. R. A. Oliver, B. Daudin, *Phil. Mag.* **87**, 1967 (2007).
34. N. K. van der Laak, R. A. Oliver, M. J. Kappers, C. J. Humphreys, *Appl. Phys. Lett.* **90**, 121911 (2007).
35. N. K. van der Laak, R. A. Oliver, M. J. Kappers, C. J. Humphreys, *J. Appl. Phys.* **102**, 013513 (2007).
36. F. C. Frank, *Acta Crystallogr.* **4**, 497 (1951).
37. R. Datta, M. J. Kappers, J. S. Barnard, C. J. Humphreys, *Appl. Phys. Lett.* **85**, 3411 (2004).
38. R. A. Oliver, S. E. Bennett, T. Zhu, D. J. Beesley, M. J. Kappers, D. W. Saxey, A. Cerezo, C. J. Humphreys, *J. Phys. D*, submitted (2010).
39. T. F. Kelly, M. K. Miller, *Rev. Sci. Instrum.* **78**, 031101 (2007).
40. M. J. Galtrey, R. A. Oliver, M. J. Kappers, C. J. Humphreys, D. J. Stokes, P. H. Clifton, A. Cerezo, *Appl. Phys. Lett.* **90**, 061903 (2007).
41. J. Narayan, H. Wang, J. Ye, S.-J. Hon, K. Fox, J. C. Chen, H. K. Choi, J. C. C. Fan, *Appl. Phys. Lett.* **81**, 841 (2002).
42. Y. Chen, H. Wenhao, *Meas. Sci. Technol.* **15**, 2005 (2004).
43. R. A. Oliver, J. Sumner, M. J. Kappers, C. J. Humphreys, *J. Appl. Phys.* **106**, 054319 (2009).
44. W. M. Tong, R. S. Williams, *Annu. Rev. Phys. Chem.* **45**, 401 (1994).
45. C. Herring, *J. Appl. Phys.* **21**, 301 (1950).
46. Y. Huang, C. C. Williams, H. Smith, *J. Vac. Sci. Technol. B* **14**, 433 (1996).
47. J. Sumner, R. A. Oliver, M. J. Kappers, C. J. Humphreys, *J. Vac. Sci. Technol. B* **26**, 611 (2008).
48. R. Stephenson, A. Verhulst, P. De Wolf, M. Caymax, W. Vandervorst, *Appl. Phys. Lett.* **73**, 2597 (1998).
49. J. Sumner, R. A. Oliver, M. J. Kappers, C. J. Humphreys, *Phys. Stat. Sol. (c)* **4**, 2576 (2007).
50. W. Lee, J.-H. Ryou, D. Yoo, J. Limb, R. D. Dupuis, D. Hanser, E. Preble, N. M. Williams, K. Evans, *Appl. Phys. Lett.* **90**, 093509 (2007).
51. S. Das Bakshi, J. Sumner, M. J. Kappers, R. A. Oliver, *J. Cryst. Growth.* **311**, 232 (2009).
52. R. A. Oliver, S. E. Bennett, J. Sumner, M. J. Kappers, C. J. Humphreys, *J. Phys.: Conf. Ser.* **209**, 012049 (2010).
53. J. Sumner, R. A. Oliver, M. J. Kappers, C. J. Humphreys, *J. Appl. Phys.* **106**, 104503 (2009).
54. O. Bowallius, S. Anand, M. Hammar, S. Nilsson, G. Landgren, *Appl. Surf. Sci.* **145**, 137 (1999).
55. M. W. Xu, T. Hantschel, W. Vandervorst, *Appl. Phys. Lett.* **81**, 177 (2002).

56. R. J. Colton, Nanoscale measurements and manipulation, *J. Vac. Sci. Technol. B* **22**, 1609 (2004).
57. X. Zhou, E. T. Yu, D. I. Florescu, J. C. Ramer, D. S. Lee, S. M. Ting, E. A. Armour, *Appl. Phys. Lett.* **86**, 202113 (2005).
58. L. J. Singh, R. A. Oliver, Z. H. Barber, D. A. Eustace, D. W. McComb, S. K. Clowes, A. M. Gilbertson, F. Magnus, W. R. Branford, L. F. Cohen, L. Buckle, P. D. Buckle, T. Ashley, *J. Phys. D* **40**, 3190 (2007).
59. M. Troyon, K. Smaali, *Appl. Phys. Lett.* **90**, 212110 (2007).
60. D. Golberg, P. M. F. J. Costa, O. Lourie, M. Mitome, X. D. Bai, K. Kurashima, C. Y. Zhi, C. C. Tang, Y. Bando, *Nano Lett.* **7**, 2146 (2007).

9

STM OF SELF ASSEMBLED III–V NANOSTRUCTURES

Vaishno D. Dasika[*] and Rachel S. Goldman[*,†]
*Department of Electrical Engineering and Computer Science
University of Michigan, Ann Arbor, MI 48109-2136, USA

†Department of Materials Science and Engineering
University of Michigan, Ann Arbor, MI 48109-2136, USA

9.1.   INTRODUCTION

Over the past few decades, advances in semiconductor thin film growth have enabled the fabrication of semiconductor heterostructures with nanometer-scale precision.[1-7] Using growth techniques such as molecular beam epitaxy (MBE), it is possible to confine the dimensions of a semiconductor to <1 nm.[5-13] Through the formation of heterostructures consisting of materials with differing bandgaps, it is possible to tailor the local energy band offsets to confine or redistribute charge carriers, thereby influencing the optical and electronic properties of the materials.[9-12] Built-in strain fields can also be utilized to alter the structural properties of heterostructures, thus enhancing their optoelectronic properties.[14-17] For example, the accumulated strain in InAs on GaAs leads to the formation of self-assembled quantum dots (QDs) via the Stranski-Krastanov (S-K) growth mode transition.

Strain-induced self-assembled InAs/GaAs QDs have enabled significant advances in several devices including light emitters and detectors, nano-biological devices, field-effect transistors, and quantum computing elements.[18-24] Variations in the size and shape of the QDs significantly impact their electronic structure and luminescence properties.[14,15] Therefore, the nanometer-scale details of the uniformity of QD organization, as well as the evolution of QD shapes and sizes are critical for the development of novel applications. A variety of techniques are available for the structural investigation of interfaces in QD heterostructures, including high-resolution transmission electron microscopy (TEM), X-ray diffraction (XRD), atom-probe tomography (APT), and cross-sectional scanning tunneling microscopy (XSTM). High resolution TEM often has a lateral resolution on the order of angstroms. However, the data consist of an average over the foil thickness, which is typically on the order of 100's of Ås.[25] Furthermore, the observed contrast in the images depends on both the strain and composition, thereby requiring advanced analysis techniques and extensive modeling to obtain quantitative structural information about the sample.[26–28] XRD measurements are typically an average over length-scales greater than several microns, and for superlattices such as the ones as discussed in this chapter, require detailed modeling to determine materials properties such as composition.[29,30] Thus, direct measurements of interdiffusion and segregation lengths are generally limited by the inherent averaging which occurs in conventional characterization techniques such as TEM and XRD. In APT, a pulsed voltage on the order of kilovolts is applied to a needle-shaped sample, resulting in the emission of ions from the sample, which are then accelerated towards a detector screen.[31] The time of flight of the emitted ions is used to identify the chemical species, and thus, 3D tomographic images of the sample can be reconstructed.[32] The indium (In) composition in QDs has recently been profiled by APT.[33] However, reconstructing the 3D images requires many assumptions about the sample such as the shape and size of the apex and the temperature and electric field distributions.[31–34]

Various techniques are available to investigate the electronic states of QD heterostructures, including photoluminescence (PL) spectroscopy, capacitance-voltage (C-V) spectroscopy, and scanning tunneling spectroscopy (STS). PL spectroscopy is typically used to measure the energy difference between the confined electron and hole states in low dimensional heterostructures.[35–38] However, PL spectra correspond to a spatial average whose interpretation requires several assumptions regarding interface abruptness, alloy composition, and lateral uniformity.[39,40] Microphotoluminescence spectroscopy (μ-PL) limits the measurement area to the order of $\sim 1\,\mu$m, but measuring spectra from single QDs using this technique requires the growth of ultra-low density QDs.[41–43] In addition, due to the fast decay times of the excited electron and hole states, PL is typically most sensitive to the energy difference between the ground electron and hole states.[40] C-V spectroscopy typically involves the measurement of the differential capacitance of a $p-n$ or $p-i-n$ heterostructure as a function of the applied bias voltage. A variation in the applied bias modifies the depletion width in the heterostructure. Thus, sweeping the applied voltage moves the depletion region through the layers of the heterostructure, thereby sweeping charged carriers through layers of QDs. In the case of QDs, for certain bias values, maxima in the CV spectra, corresponding to energy levels in the QDs, are observed.[44] However, the data typically corresponds to a spatial average over 100's of microns.[45–48]

XSTM allows direct observations of the spatial distribution of individual atoms at the surface.[49–54] An important advantage of XSTM over the cleaved techniques described above is that the images are associated primarily with the top layer of the cleaved surface, instead of an average over many layers. Furthermore, the UHV-cleaved (110) surface of most III–V compounds does not reconstruct, and the dangling bond states do not lie within the energy band gap.[51] Thus, it is possible to obtain information about the bulk-like structure, chemistry, and electronic properties of the layers. To investigate the electronic states of heterostructures, scanning tunneling spectroscopy (STS), which allows spatially-resolved electronic measurements within single layers of semiconductors, is a promising alternative to methods such as PL and C-V spectroscopy.

The atomic structure of InAs/GaAs QDs has been reported to be influenced by a number of growth parameters including substrate temperature, III/V flux ratio, growth rate, the presence of alloy buffer and/or capping layers, and post-growth annealing.[3,13,55–58] For example, growth with an alternating supply of anion and cation species is expected to increase the cation diffusion length, thereby lowering the film thickness for the 2D to 3D S-K growth mode transition, resulting in larger dots.[1,2,4–7] Furthermore, intermixing between the InAs dots and the GaAs buffer and capping layers leads to alloy formation within the dots and the surrounding wetting layer (WL).[59,60] In addition, it has been reported that InAs/GaAs dots grown on InGaAs buffers have higher densities than those grown directly on GaAs.[3,13,55] However, there have been conflicting reports on the effect of alloy buffers on dot size. In one report, it was suggested that the presence of an alloy buffer does not influence dot size;[13] others suggested that similar alloy buffers lead to an increase[55] or decrease[3] in dot dimensions. In addition, growth of InAs/GaAs dots with InGaAs in lieu of GaAs capping layers apparently minimizes the tendency for the reduction in dot height upon capping, often termed dot "collapse".[56–58] However, as described above, the influence of alloy capping layers has primarily been investigated qualitatively using plan-view STM or AFM, and therefore, the quantitative structure of the buried dots was not resolved.

Several reports have suggested that QDs often have non-uniform compositions across their width and height due to indium segregation and inter-diffusion at the InAs/GaAs interfaces.[28,60–65] These reports suggest that the lateral [In] is highest at the QD core and that the vertical [In] increases in the growth direction. Between the QDs, the WL contains sparse concentrations of individual In atoms which have not agglomerated to form a 3D island.[66] The WLs are typically 2D inhomogeneous films with significant [In] gradients, including vertical In segregation and lateral In clustering.[66–71]

Doping of III–V semiconductors with transition metals such as manganese (Mn) leads to simultaneous semiconducting and ferromagnetic behavior, thus enabling devices such as spin-valves and spin-injection contacts.[72–75] In the case of QDs, Mn-doping enables the achievement of spin-polarized optoelectronic devices such as lasers and LEDs.[74,75] For epitaxially grown GaMnAs heterostructures, ferromagnetism has been reported below the Curie temperature of $\sim$180 K, although ion beam-induced MnAs nano-clusters in GaAs have led to Curie temperatures as high as 360 K.[76,77] Furthermore, for InAs:Mn QDs, Curie temperatures >300 K have been reported.[74,78] However, there have been conflicting reports on the distribution of the Mn dopants in the InAs:Mn QDs and surrounding matrix. One group reported TEM-based electron energy loss spectroscopy (EELS) measurements of large (36 ± 1 nm diameter) InAs:Mn QDs, suggesting that Mn dopants reside primarily in the QD core.[78] Another group showed XSTM evidence of significant Mn surface segregation during InAs:Mn QD growth, and suggested that the majority of the Mn dopants in small (<20 nm diameter) QDs are located at the edges of the QD and/or in the GaAs matrix outside the QD.[79]

In this chapter, we describe XSTM and STS and demonstrate several applications of XSTM and STS to investigate QD heterostructures, from the Ph.D. thesis of Dasika.[80] For example, we discuss the application of XSTM to study the effects of alloyed buffer and capping layers on the QDs and the WL.[81] In addition, we present an example of the application of XSTM and STS to explore the origins of the effective bandgap variations in individual (uncoupled) QDs and the surrounding WL, by investigating the influence of compositional variations on the electronic states of QDs and WLs.[82] Furthermore, we discuss an analysis of the distribution of Mn dopants and their influence on the electronic states of InAs:Mn QDs and the surrounding GaAs matrix using a combination of XSTM and STS.[83]

This chapter is organized as follows. In Section 2, XSTM and STS are described in detail. In Section 3, details regarding heterostructure design for successful XSTM experiments, and the growth of the samples discussed in this chapter are provided. In Section 4, we review our investigations of the influence of InGaAs alloy buffer and capping layers on the size, shape, and density of InAs/GaAs QDs and corresponding WLs. Large-scale and high-resolution XSTM images enabled measurement of the dimensions, density, and WL thicknesses for QDs with and without surrounding InGaAs layers.[81] In Section 5, we use a combination of XSTM and STS to explore the origins of the effective bandgap variations in individual, uncoupled QDs and the surrounding WL.[82] In Section 6, we present an analysis of the atomic-scale distribution of Mn dopants and its influence on the electronic states of InAs:Mn QDs and the surrounding GaAs matrix, using XSTM[84] and STS,[82] in comparison with order(N) tight-binding calculations of the local density of states (LDOS).[83,85] A summary is provided in Section 7.

9.2.   CROSS-SECTIONAL SCANNING TUNNELING MICROSCOPY (XSTM) AND SCANNING TUNNELING SPECTROSCOPY (STS)

9.2.1.   Cross-sectional Scanning Tunneling Microscopy (XSTM)

For XSTM, a cross-section of a multilayer sample cleaved in ultra-high vacuum (UHV), to expose an atomically flat surface. Constant-current STM is then performed on the exposed surface, as illustrated in

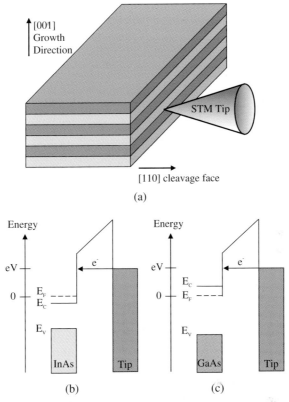

Fig. 9.1   Schematic of cross-sectional scanning tunneling microscopy applied to III-V heterostructures. Typically, after cleaving, constant-current STM is performed on the cleaved
surface, shown in (a). Schematics of the tunneling process between a tip and a sample under
positive sample bias are shown in (b) and (c). The electrons tunnel from the tip into the
energy levels above the Fermi level (E_F) of either (b) InAs or (c) GaAs. Since there are more
states available to tunnel into in InAs than GaAs, the InAs layers are expected to appear
brighter in a constant-current STM image. Adapted from Ref. 80.

Fig. 9.1.[80] When the cleaved surface is atomically flat, with monolayer
steps spaced hundreds of nanometers apart, the apparent topographic
contrast observed in constant-current images is primarily due to variations in the electronic properties of the individual layers.[49]

   An example of an XSTM topographic image is shown in
Fig. 9.2(a).[80] The bright and dark regions visible in the image correspond to layers of InAs and GaAs, respectively. Figures 9.1(b) and 9.1(c)
show schematic energy band diagrams for empty state imaging of the

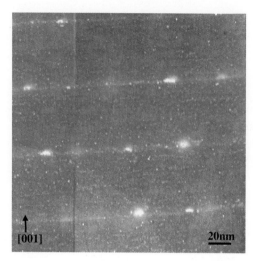

Fig. 9.2   (a) XSTM topographical image of InAs dots in a GaAs matrix. The apparent topographic contrast is due to differences in the density of states of the InAs and GaAs. Reprinted from Ref. 80.

InAs and GaAs, respectively. In both cases, the application of a positive sample bias voltage, V, results in electron tunneling from the STM tip into the empty conduction band states of the semiconductor. Since the bandgap of GaAs (Fig. 9.1(c)) is larger than the bandgap of InAs (Fig. 9.1(b)), fewer states are available for the electrons to tunnel into the GaAs than into the InAs. Thus, the STM tip must move closer to (away from) the GaAs (InAs) surface to maintain a constant tunneling current. Therefore, the GaAs (InAs) layer appears darker (brighter) in the XSTM image.

To differentiate layers in constant-current XSTM images, we consider lateral variations in the tip height.[54] For example, to differentiate the GaAs, the QDs, and the clustered regions of the WL for all studies described in this chapter, we estimated the tip height criterion as follows. Bright regions with maximum tip heights at least 2.1 Å above the GaAs background were considered to be possible QDs. Within the bright regions, pixels with tip heights at least 1.1 Å above the GaAs background were considered to be part of the QD.[54] Surrounding the QDs, bright regions with maximum tip heights between 0.4 and 1.5 Å above the GaAs background were considered to be possible clustered WL In atoms.

Within these regions of the WL, we estimated an indium atom tip height criterion of 0.85 Å ± 0.05 Å with respect to the GaAs background.[86]

9.2.2. Scanning Tunneling Spectroscopy (STS)

It is also possible to obtain spectroscopic information about semiconductor heterostructures using conductance imaging and variable tip-sample separation spectroscopy (VS-STS).[87,88] In the VS-STS method, both the bias voltage and tip-sample separation are varied while the tunneling current and differential conductance (dI/dV) are measured. Specifically, the feedback loop is deactivated, and a continuous linear ramp of the sample bias voltage is applied, while the tip height is varied in a controlled manner.[89] The tip is moved towards (away from) the surface as the magnitude of the bias voltage is decreased (increased). As a result, the measured tunneling current and differential conductance are increased in the vicinity of the band edges, enabling an accurate determination of their energetic positions. The main advantage of the VS-STS method is that the conductance and current at low voltages are amplified while the noise level remains constant. The dynamic range, i.e. the ratio of the largest to smallest detectable signal, increases by 2–3 orders of magnitude in comparison with that of constant-separation STS, thus enabling accurate identification of the energetic positions of the band edges.[90]

9.2.3. Variable Separation STS: Experimental Steps and Analysis

The first step of VS-STS involves setting up the appropriate tip extension, $z(V)$, as shown in Fig. 9.3(a).[54] Care must be taken to avoid damaging the tip and sample by extending the tip too much. Typically, the tip is moved 6–10 Å toward (away from) the sample while the applied bias voltage is increased from −2.5 to 0 V (increased from 0 V to 2.5 V). The precise parameters are optimized for each particular tip-sample combination.

The measured sample current as a function of voltage, $I(V)$, is plotted in Fig. 9.3(b). The measured current is positive (negative) for large positive (negative) voltage values, while the current is negligible for voltage values in the vicinity of 0 V. The sample voltage corresponds to

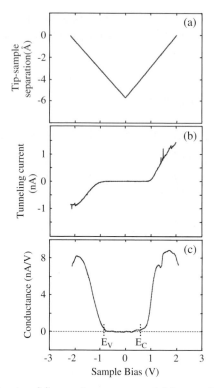

Fig. 9.3 STS data showing (b) tunneling current and (c) normalized conductance as a function of sample bias, obtained from a cleaved GaAs (110) surface. The variation in tip-sample separation, shown in (a), represents the motion of the tip towards the surface from −2.2 to 0 V, and away from the surface from 0 to +2.2 V. For plot (c), the boundaries between regions of positive conductance and regions of negligible conductance correspond approximately to the valence and conduction band edges GaAs. From Ref. 54.

the energy of the state relative to the Fermi level. Thus, 0 V corresponds to the Fermi level.

The differential conductance, dI/dV is measured using a lock-in technique,[80] and is plotted as a function of sample bias in Fig. 9.3(c). The plot consists of three regions: a region of increased differential conductance for voltage values less than −1 V, followed by a region of negligible conductance for low absolute voltages, and finally, another region of increased conductance for voltage values greater than 1 V. The boundaries of these regions correspond approximately to the valence (E_V) and conduction band edges (E_C) of the GaAs, respectively.

Prior to the collection of STS spectra from a region of interest such as the QDs, calibration spectra on "known" regions, such as GaAs, are first collected. In a region of GaAs, the tip is sequentially moved closer to the sample in steps of 1–2 Å, until a reasonable bandgap $(1.43 \pm 0.5\,\text{eV})$ is measured for the known layers.

9.2.4. Tip-Induced Band Bending

In GaAs and related semiconductors, tip-induced band bending (TIBB), or dynamic band bending, can introduce shifts in the measured band edges ranging from several tenths of an eV (for doping concentrations on the order of $10^{17}\,\text{cm}^{-3}$) to $\sim 0.1\,\text{eV}$ (for doping concentrations greater than $10^{18}\,\text{cm}^{-3}$).[90] TIBB occurs when the applied voltage bias is dropped across both the vacuum gap and the semiconductor itself, producing shifts in the measured band edges.[90–95] TIBB occurs when the tip and the semiconductor have significantly different Fermi levels. To maintain thermal equilibrium, a redistribution of surface charges takes place in the semiconductor, thus producing a depletion region a the surface of the semiconductor, as shown in Fig. 9.4(a).[80]

When a voltage is applied, significant band bending can take place within the semiconductor, as shown in Fig. 9.4(b) and Fig. 9.4(c), where the apparent conduction and valence band edges due to TIBB are labeled "E'_C" and "E'_V" respectively, while the predicted band edges are labeled "E_C" and "E_V". When the applied bias is negative, the Fermi level of the tip, E_{Ft}, is lower in energy than the Fermi level of the sample, E_{Fs}, and electrons tunnel from the valence band of the sample to the tip, as shown in Fig. 9.4(b). However, due to TIBB, the apparent valence band edge, E'_V is lower than the predicted band edge, E_V. Similarly, when the applied bias is positive, $E_{Ft} > E_{Fs}$ and electrons tunnel from the tip to the conduction band of the sample, as shown in Fig. 9.4(c). Due to TIBB, the apparent conduction band edge, E'_C, is higher than the predicted band edge, E_C. For an STS measurement with the voltage sweep from negative to positive values, the apparent band gaps are shifted so that the measured band gap, $E'_C - E'_V$ is larger than the expected band gap, $E_C - E_V$.

Thus, in GaAs and related semiconductors, TIBB can introduce shifts in the measured band edges ranging from several tenths of an eV (for doping concentrations on the order of $10^{17}\,\text{cm}^{-3}$) to $\sim 0.1\,\text{eV}$ (for doping concentrations greater than $10^{18}\,\text{cm}^{-3}$).[90] Some of the methods

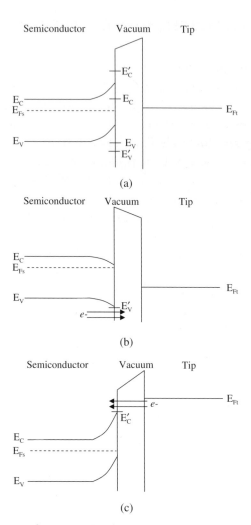

Fig. 9.4   Band diagram of n-type semiconductor-vacuum-tip tunnel junction at (a) zero bias (b) positive bias, and (c) negative bias, with respect to the sample. E_{Fs} (E_{Ft}) corresponds to the Fermi level of the sample (tip). Due to tip-induced band bending, the measured positions of the conduction (E_C') and valence (E_V') band edges are shifted relative to the expected positions of the conduction (E_C) and valence (E_V) band edges. Reprinted from Ref. 80.

to minimize the effects of TIBB are heavily doping the sample, and passivating the cleaved surface. Both methods pin the surface Fermi level, thereby preventing the formation of a depletion region in the semiconductor.[91,96] Typically, the sample is heavily doped.

9.3. DESIGN OF HETEROSTRUCTURES FOR XSTM

Several heterostructure design strategies are often used to optimize the XSTM experiment success rate. These include the design of a strain balanced structure, the use of marker layers, and the use of high doping levels in the substrate and layers of interest. To increase the probability of obtaining an atomically flat cleave in the region of interest, the heterostructures are designed to be strain-balanced, with a typical cap layer thickness ≥ 300 nm. In addition, to enable the identification of a given region of the sample, marker layers are incorporated into the structure. For example, for the InAs/GaAs QD heterostructures described in this chapter, AlAs/GaAs superlattices sandwiched between layers of GaAs provided electronic contrast with respect to the surrounding GaAs, thereby functioning as a marker layer. To increase the sample-tip tunneling current and to reduce the effects of TIBB, as described in the previous section, high doping concentrations within the substrate and heterostructure layers are needed. Thus, samples for XSTM are typically highly doped ($>10^{18}$ cm^{-3}), which, for n-type GaAs, corresponds to a resistivity less than $10^{-3} \Omega$-cm. Typical doping concentrations and resitivities for GaAs doped with standard dopants Si (n-type) or Be (p-type) are provided in Table 9.1.

The samples in Sections 9.4 and 9.5 were grown via an alternate supply of anion and cation species, often termed migration enhanced epitaxy (MEE).[97] The heterostructures consisted of one layer of InAs dots with In$_{0.2}$Ga$_{0.8}$As alloy buffer and capping layers (termed alloy quantum dots or "AQDs"), and 3 sets of SLs with 1, 3, and 8 layers of 3 ML

Table 9.1. Typical doping concentrations and resitivities of p-type and n-type GaAs.

	p-type GaAs (Dopant: Be)	n-type GaAs (Dopant: Si)
Doping concentration (cm^{-3})	[115]1×10^{19} [116,117]5×10^{18}	[115]5×10^{18} [118]3×10^{18} [117]1×10^{18} [119]4×10^{17}
Resistivity (Ω-cm)	[115]0.0016 [117]0.0031	[115]0.00015 [118]0.00045 [117]0.00070 [119]0.0018

InAs QDs, without surrounding alloy layers, followed by 50 nm of GaAs. Each set of SLs was separated by a multilayer consisting of 20 or 40 nm of AlAs/GaAs short-period SLs sandwiched between two 70 nm GaAs layers. The QD and AQD layers were separated by >50 nm of GaAs to prevent dot stacking and coupling. In both cases, for dot growth, the In and As were deposited alternately for 8s, followed by a 5s growth interruption in the absence of As to allow dot nucleation.[97] The AQDs were grown on 1.25 nm of $In_{0.2}Ga_{0.8}As$, followed by capping with 7.5 nm of $In_{0.2}Ga_{0.8}As$. The entire structure was grown at 480°C and capped with 300 nm GaAs. All layers, except the InAs and $In_{0.2}Ga_{0.8}As$ layers, were Si-doped at $\sim 3 \times 10^{18}$ cm^{-3}.

For the samples discussed in Section 9.6, heterostructures consisting of InAs:Be QDs and InAs:Mn QDs, separated by 20 nm of GaAs, were grown on (001)-oriented $p+$ GaAs.[78] Using molecular-beam epitaxy, a 500 nm GaAs buffer layer, followed by 40 nm AlAs/GaAs superlattices and 50 nm of GaAs, were grown at 610°C. The substrate temperature was then reduced to 280°C, and 4.2 ML of InAs were deposited, followed by a 45 second growth interruption to allow quantum dot (QD) formation. Following a 20 nm GaAs spacer layer, the InAs:Mn QDs were grown. For the InAs:Mn QDs, 2.0 monolayers (MLs) of InAs were initially deposited in the absence of manganese (Mn). The Mn effusion cell shutter was then opened for the growth of the final 2.2 ML of InAs, thus producing InAs:Mn QDs. QD growth was again followed by a 45 second growth interruption prior to additional capping at the low growth temperature. During the growth of this structure, As_4:Ga and In:Mn beam equivalent pressure (BEP) ratios of $\sim$16 and 20 were used, and the InAs growth rate was 0.07 ML/s. The structure was capped with 500 nm GaAs, and all layers, except the InAs:Mn QDs, were Be doped at $\sim 5 \times 10^{18}$ cm^{-3}.

9.4.   INFLUENCE OF ALLOY LAYERS ON QD SIZE AND DISTRIBUTION

Here, we describe investigations of the influence of InGaAs alloy buffer and capping layers on the size, shape, and density of InAs/GaAs dots and corresponding WLs.[81] We compared QDs, i.e. dots with GaAs buffer and capping layers, and dots with alloy buffer and capping layers, termed alloy quantum dots or "AQDs". As described in Ref. 81, large-scale and high-resolution XSTM images reveal larger dimensions, density, and WL thicknesses for the AQDs in comparison with the QDs. Taking into account the reduction in misfit strain provided by the InGaAs alloy layers at the buffer/dot and dot/cap interfaces, we describe a strain-based mechanism for QD formation and collapse in the absence and presence of InGaAs alloy layers. This mechanism is likely to be applicable to a wide range of lattice-mismatched thin-film systems.

9.4.1. QD Dimensions, Density, and WL Thickness

Figure 9.5 shows example large-scale XSTM topographic images of the (a) QDs and (b) AQDs.[81] In Fig. 9.5(a), the bright ellipses surrounded by darker layers correspond to InAs QDs in GaAs. Due to the relatively thick spacer layer between the QD and AQD layers, the dots are uncorrelated, consistent with earlier reports.[98] Furthermore, the WL between the QDs contains regions of In clustering, as indicated in Fig. 9.5(a), similar to earlier XSTM observations of MBE-grown InAs/GaAs QDs.[69] The bright region in Fig. 9.5(b) corresponds to a layer of AQDs. Based upon Fig. 9.5, and several similar images, it is apparent that the dot sizes and the WL thicknesses are higher for the AQDs in comparison with the QDs. It is likely that intermixing with the surrounding GaAs diminished the QD dimensions and WL thickness whereas intermixing with the $In_{0.2}Ga_{0.8}As$ alloy layers both above and below the AQDs, lead to an increase in the AQD dimensions and WL thickness.

To quantify the influence of the surrounding $In_{0.2}Ga_{0.8}As$ alloy layers on dot dimensions and densities, we examined several high resolution images of the QD and AQD layers, spanning $>0.5\,\mu m^2$, and applied the line-cut analysis described in Section 2. The average QD and AQD dimensions (width, height, WL thickness) are plotted in Fig. 9.6.[81] The average QD diameter and height are 16 ± 3 nm and 7 ± 1 nm, respectively, while the AQDs are larger, with average diameters and heights of 22 ± 3 nm and 12 ± 2 nm, respectively. Thus, the diameters (heights) of the AQDs are 38% (71%) greater than those of the QDs. In addition, the WL between the QDs is 2.0 ± 0.8 nm thick whereas the WL between

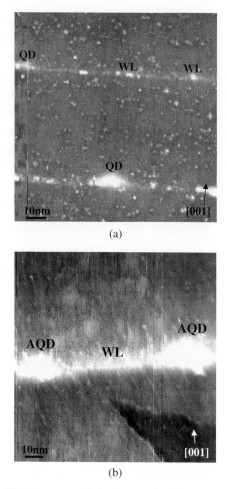

(a)

(b)

Fig. 9.5   Large-scale XSTM topographic images acquired at a sample bias of $-2.0\,$V, with bright regions corresponding to InAs. (a) QDs: 3 ML InAs dots in a GaAs matrix. The gray-scale range displayed is 0.7 nm. (b) AQDs: 3 ML InAs dots grown between a 1.25 nm of $In_{0.2}Ga_{0.8}As$ buffer, and a 7.5 nm of $In_{0.2}Ga_{0.8}As$ capping layer. The gray-scale range displayed is 0.8 nm. The dot dimensions and WL thickness are greater for the AQDs, in comparison to that of the QDs. We note that the observed periodicity is due to interference from a lock-in amplifier during image acquisition. Reprinted with permission from Ref. 81. Copyright 2009, American Institute of Physics.

the AQDs is significantly thicker at 8 ± 2 nm. To infer dot densities, we also determined the average lateral spacing between the QDs and AQDs. The average lateral spacing between the QDs is 80 ± 21 nm whereas the average lateral spacing between the AQDs is 54 ± 12 nm, suggesting that

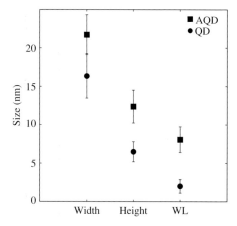

Fig. 9.6 Comparison of width, height, and WL thickness for the QDs and AQDs. These values are based on several high resolution images of the QD and AQD layers (such as those shown in Fig. 9.5), spanning $>0.5\,\mu m^2$. Reprinted with permission from Ref. 81. Copyright 2009, American Institute of Physics.

the alloy buffer layer promoted an increase in dot density. Thus, use of the alloy layers as both buffer and capping layers apparently leads to an increase in WL thickness and an increase in dot width, height, and density.

9.4.2. Mechanism for Dot Formation and Collapse

In Fig. 9.7, we present a mechanism proposed for dot formation in the absence and presence of InGaAs alloy buffer layers.[81] The diagrams on the left (right) represent the growth stages and associated surface strain for the QDs (AQDs), including buffer layer growth (Fig. 9.7(a) and (b)), initial InAs deposition for dot formation (Fig. 9.7(c) and (d)), and dot growth (Fig. 9.7(e) and (f)). Initially, the "substrate" for QDs is an unstrained GaAs buffer (Fig. 9.7(a)), and the "substrate" for AQDs is the strained $In_{0.2}Ga_{0.8}As$ alloy (Fig. 9.7(b)). During the growth of the alloy buffer layer, the In diffuses laterally and segregates vertically, forming regions of varying [In] and surface strain on the growth surface, as illustrated in Fig. 9.7(b).[99] For QD growth directly on GaAs, shown in Fig. 9.7(c), a 2D WL, with strain distribution similar to that in Fig. 9.7(b), is observed initially. The targeted buffer/dot misfit,

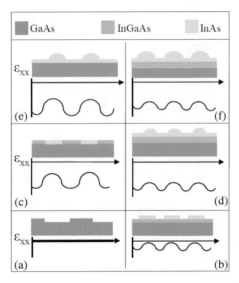

Fig. 9.7   Mechanism for dot formation in the absence and presence of InGaAs alloy layers, Part I: buffer layer growth prior to InAs deposition for the (a) QDs and (b) AQDs; initial stages of InAs deposition for the (c) QD layers and (d) AQD layers; dot nucleation for the (e) QD layers and (f) AQD layers. Reprinted with permission from Ref. 81. Copyright 2009, American Institute of Physics.

$\varepsilon_f = (a_f - a_s)/a_s$, where a_f and a_s are the film and substrate lattice constants, is 7.2% (5.7%) for the QDs (AQDs). Beyond a critical thickness, proportional to $1/\varepsilon_f^2$, the WL becomes unstable to surface perturbations, allowing the formation of islands.[100,101] Since the buffer/dot ε_f is lower for the AQDs, the AQD critical thickness is expected to be higher than the QD critical thickness.[101] However, the alloy buffer layer acts as a "pre-existing" WL, so that when InAs is deposited on the alloy buffer, the In atoms nucleate to form dots in regions where the [In] is higher, as shown in Fig. 9.7(d). Therefore, the AQD density is initially higher than the QD density. As more InAs is deposited, QD nucleation occurs, and the QDs increase in size, as shown in Fig. 9.7(e). The compressive strain in the InAs layer is partially elastically relaxed by QD formation, leading to the lateral surface strain variation shown in Fig. 9.7(e), where the edges of the QD are under higher compressive strain than the top surface of the QD. With additional InAs deposition, the AQDs also increase in size, as shown in Fig. 9.7(f). However, due to the lower buffer/dot

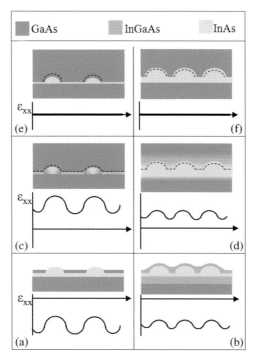

Fig. 9.8   Mechanism for dot formation in the absence and presence of InGaAs alloy layers, Part II: initial stages of capping with (a) GaAs for the QDs and (b) $In_{0.2}Ga_{0.8}As$ for the AQDs; additional capping and intermixing for the (c) QDs and (d) AQDs, and final capped structures of the (e) QDs and (f) AQDs. The dotted line represents the dot height prior to capping. Reprinted with permission from Ref. 81. Copyright 2009, American Institute of Physics.

misfit strain, the compressive strain on the surface is lower for the AQDs than for the QDs.

In Fig. 9.8, we present a mechanism proposed for dot collapse in the absence and presence of InGaAs alloy capping layers.[81] The diagrams on the left (right) represent the growth stages and associated surface strain for the QDs (AQDs), including the early stages of capping (Fig. 9.8(a) and (b)), intermixing after additional capping (Fig. 9.8(c) and (d)), and the final dot structures after capping (Fig. 9.8(e) and (f)). As the QDs are capped with GaAs, Ga preferentially accumulates in regions of highest compressive strain such as the QD edges as shown in Fig. 9.8(a). With increasing cap thickness, the high compressive dot/cap ε_f at the QD base facilitates the diffusion of In atoms away from the

QD, leading to QD collapse, as shown in Fig. 9.8(c).[56] On the other hand, as the AQDs are capped with $In_{0.2}Ga_{0.8}As$, In adatoms preferentially attach at the regions of the lowest compressive strain, namely on the top surface of the AQD, as shown in Fig. 9.8(b). Thus, $In_{0.2}Ga_{0.8}As$ accumulates both on top of and at the edges of the AQDs. The In from the surrounding alloy layers diffuses into the AQDs and WL, leading to an increase in AQD dimensions and WL thickness, as illustrated in Fig. 9.8(d). Thus, the increase in QD size and WL thickness is limited, as shown in Fig. 9.8(e), whereas the surrounding alloy layers promote an increase in AQD and WL dimensions, as shown in Fig. 9.8(f).

9.5. QUANTUM DOT AND WETTING LAYER ELECTRONIC STATES

Here, we describe an XSTM and STS investigation of electronic states in nanostructures, in which the lateral and vertical variations of the effective bandgap within individual QDs and clustered regions of the WL are revealed.[82] For both the QD and WL regions, the effective bandgap is likely the energy difference between the lowest confined electron (E_e) and hole (E_h) energies. We examined several high resolution images of the QDs spanning $>0.5\,\mu m^2$ and acquired STS spectra from more than 70 QDs. We find decreases in the effective bandgap both laterally, towards the QD core, and vertically, in the growth direction. These trends are consistent with an increase in indium concentration, [In], toward the center and top of the QD. Similarly, in the clustered regions of the WL, the effective bandgap variations are dominated by variations in the [In].

9.5.1. Uncoupled QD Electronic States

An example large-scale XSTM topographic image of the QDs is shown in Fig. 9.9, where the bright ellipses surrounded by darker layers correspond to InAs QDs in GaAs.[82] Due to the relatively thick (50 nm) spacer layer between QDs, the QDs are uncorrelated. Furthermore, the WL appears discontinuous and contains regions of significant In clustering. To differentiate the GaAs, the QDs, and the clustered regions of the WL, we estimated a tip height criterion, as described in Section 2. STS spectra

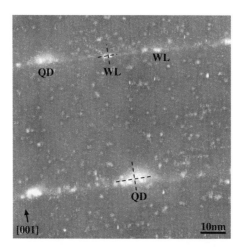

Fig. 9.9   Large-scale XSTM topographic image of the uncoupled InAs/GaAs QDs, with bright regions corresponding to InAs. WL regions with significant In clustering are labeled. Spatially resolved STS spectra were acquired both laterally and vertically from the QDs and the clustered regions of the WL, as indicated by the dashed lines near the top and bottom of the image. The image was acquired at a sample bias of −2.0 V, and the gray-scale range displayed is 0.7 nm. Reprinted with permission from Ref. 82. Copyright 2009, American Institute of Physics.

were then acquired both laterally and vertically across the QDs and the clustered regions of the WL, as indicated by the dashed lines in Fig. 9.9.

Figure 9.10(a) shows an example XSTM image where the bright ellipse, with major and minor axes of 16 nm and 6 nm respectively, corresponds to an InAs QD in a GaAs matrix.[82] In Fig. 9.10(b), the normalized conductance versus sample bias voltage is plotted for the edge and center of the QD shown in Fig. 9.10(a), in comparison with a region of clean GaAs. The GaAs spectrum, shown at the bottom of Fig. 9.10(b), displays well-defined band edges, with a bandgap of 1.45 eV, similar to that of bulk GaAs at room temperature.[102] In Fig. 9.10(b), at the edge of the QD, the effective bandgap is 1.09 eV (plot 1), while at the QD core, the effective bandgap is 0.87 eV (plot 2). Thus, a gradient in the effective bandgap is apparent, with the effective bandgap decreasing laterally toward the QD core. A similar trend was observed in a real-space computational study using a moments-based tight-binding method,[85] in which atomic resolution XSTM images of QDs were used to determine the position and types of atoms making up the image.[103] The

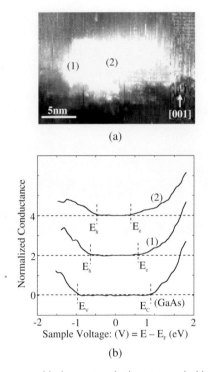

(a)

(b)

Fig. 9.10   (a) XSTM topographic image acquired at a sample bias of −2.0 V. The gray-scale range displayed is 1 nm. In (b), spatially resolved STS spectra from points (1) and (2) are plotted in comparison with a region of clean GaAs. The lowest confined hole (E_h) and electron level (E_e) are indicated by vertical dashed lines at negative and positive sample voltages, respectively. The sample voltage corresponds to the energy relative to the Fermi level. Reprinted with permission from Ref. 82. Copyright 2009, American Institute of Physics.

computations suggested a variation in the effective bandgap within the QD, with the narrowest bandgap near the QD center. However, in that study, the QD was assumed to consist of pure InAs, and only lateral variations in effective bandgap were considered. To gain a more thorough understanding of the effective bandgap variation within and between QDs, we measured lateral and vertical variations in effective bandgap and compared the observed trends to predictions of effective bandgap variations due to variations strain, shape, and composition.

Spatially-resolved STS spectra were collected from several ellipse-shaped QDs with 14 ± 1 nm major axes. The plots of normalized conductance vs. energy, such as that shown in Fig. 9.10(b), were then

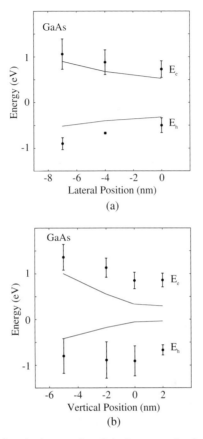

Fig. 9.11   Spatial variations in the energies of the lowest confined electron (E_e) and lowest confined hole (E_h) state in the GaAs and QD in the (a) horizontal and (b) vertical directions, with respect to the QD center. The calculated $In_xGa_{1-x}As$ band edges with $x = 0.35$ at the QD edge, $x = 0.65$ at the QD core, $x = 0.6$ at the QD bottom, and $x = 0.9$ at the QD top surface are indicated by the solid line. The x values were determined from measurements of [In] across similarly-sized QDs.[62] Reprinted with permission from Ref. 82. Copyright 2009, American Institute of Physics.

used to determine the energetic positions of the effective valence and conduction band edges, which presumably correspond to E_h and E_e. In Fig. 9.11(a) and Fig. 9.11(b), average values of E_h and E_e as a function of lateral position within a QD are plotted as solid circles.[82] For comparison, the solid line in the plot is an estimate of the lateral band-edge variation for an undoped bulk-like InGaAs alloy (without quantum confinement), based upon literature reports for In composition gradients for similar-sized QDs.[62] In a QD, E_e and E_h are expected to be higher in energy than the band edges of a bulk semiconductor. Thus, the differences between the measured E_e and E_h, and the InGaAs band-edge variation described above are likely due to a combination of carrier confinement plus TIBB.

The lateral variations in QD effective bandgap are shown in Fig. 9.11(a). In this case, the effective bandgap is narrowest at the QD center. Since the InAs/GaAs misfit strain is predicted to increase the lateral effective bandgap towards the QD core,[15] it is unlikely that strain is dominating the lateral effective bandgap variation. XSTM measurements and calculations of the In distribution and alloying in QDs have suggested an enrichment of In toward the QD core. Therefore, it is possible that lateral variations in the [In] in the QD are dominating the lateral effective bandgap variation.[61,62,105]

The vertical variations in the QD effective bandgap are shown in Fig. 9.11(b). Here, the effective bandgap is narrowest at the top of the QD. Since the QDs are wider at the center and the InAs/GaAs misfit strain is predicted to increase the effective bandgap in the growth direction,[15,17] it is unlikely that strain or vertical variations in the well width are dominating the effective bandgap variation in the growth direction. Since both the lateral and vertical variations of E_e and E_h follow the trend of the In-composition gradient induced band-edge variations, it is likely that [In] variations dominate both the lateral and vertical effective bandgap variations. A similar gradient in the effective bandgap of InAs/GaAs QDs was reported in another study.[104]

9.5.2.   Wetting Layer Electronic States

STS spectra were collected both laterally and vertically across WL regions with significant In clustering.[82] The normalized conductance was used to estimate the energetic positions of E_e and E_h. The resulting effective bandgap variation as a function of position within the WL laterally and vertically is shown in Fig. 9.12(a) and (b) respectively.[82] In the plot, '0' marks the edge of a clustered region. For comparison, the solid line shows the band edges for bulk $In_xGa_{1-x}As$, determined using x values from earlier XSTM studies of the WL.[67,69] Laterally, the

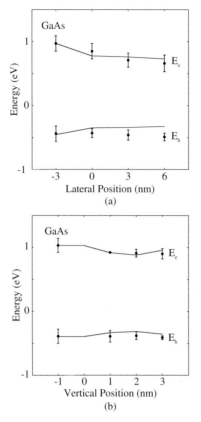

Fig. 9.12 Spatial variations in the energies of the lowest confined electron (E_e) and lowest confined hole (E_h) state in the WL (a) laterally and (b) vertically with respect to the WL edge. The calculated $In_xGa_{1-x}As$ band edges with $x = 0.26$ at the edge of the WL clustered region, $x = 0.32$ at center, $x = 0.15$ at the bottom of the clustered region, and $x = 0.10$ at the top surface is indicated by the solid line. The x values were determined from measurements of [In] across the WL between QDs.[67,69] Reprinted with permission from Ref. 82. Copyright 2009, American Institute of Physics.

effective bandgap decreases towards the center of the cluster, where the [In] is presumably the highest, similar to the trend observed in the QDs, discussed above. Vertically, there is limited variation in the [In] and correspondingly, the effective bandgap is nearly constant, $1.31 \pm 0.01\,eV$. Thus, as discussed in Section 9.4 of this chapter, the [In] in the WL varies due to interdiffusion and surface segregation, and as described in this section, the [In] gradients dominate the effective bandgap in the WL and the QDs.

9.6. INFLUENCE OF Mn DOPANTS ON InAs/GaAs QUANTUM DOT ELECTRONIC STATES

The combination of XSTM and STS is also useful for examining the incorporation of dopant atoms into QDs. Here, we review an analysis of the atomic-scale distribution of Mn dopants and their influence on the electronic states of InAs:Mn QDs and the surrounding GaAs matrix, using XSTM and STS, in comparison with order(N) tight-binding calculations of the local density of states (LDOS).[83,85] This combined approach allows us to deduce the positions of the dopants based on their influence on the LDOS. STS spectra from various locations in InAs and InAs:Mn QDs were compared with the LDOS of InAs and InAs:Mn QDs, calculated with Mn at the QD core, QD edge, or in the GaAs matrix. The observed trends in LDOS match those of the measured STS spectra *only* for the case of Mn at the QD edge, and not for the cases of Mn at the QD core or in the GaAs matrix. Together, these results suggest that the concentration of Mn dopants increases towards the edges of the QD, consistent with the predictions of a thermodynamic model for the nanoscale-size dependence of dopant incorporation in nanostructures.[105]

9.6.1. Doped QD Electronic States: XSTM and STS

Figure 9.13 shows an example of a large-scale XSTM image of the InAs:Be and InAs:Mn QDs. The bright regions towards the top (bottom) of the image correspond to the InAs:Mn (InAs:Be) QDs. Monolayer height steps are also visible towards the top of Fig. 9.13. The typical InAs:Mn QD diameters are 12 ± 2 nm and typical QD heights are 4 ± 1 nm, while the average lateral spacing between the QDs is 87 ± 30 nm. Thus, the doped QDs have smaller sizes and larger lateral spacings than undoped QDs.[106] The size difference is likely due to the surfactant nature of Mn, which would lead to smaller, lower density QDs.[107,108]

Figure 9.14 presents the STS spectra collected from the (a) QD core, (b) QD edge and (c) GaAs matrix to compare the InAs, the InAs:Be, and the InAs:Mn QDs. For each spectrum, E_h and E_e are indicated by vertical dashed lines at negative and positive sample voltages, respectively. The

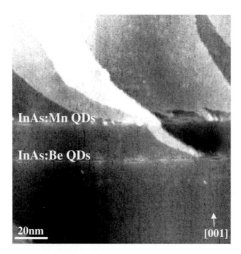

Fig. 9.13   Large-scale XSTM topographic images, with bright regions corresponding to the QDs. The bright regions towards the top (bottom) of the image correspond to the InAs:Mn (InAs:Be) QDs. The image was acquired with a constant tunneling current of 0.15 nA and a sample bias of −2.0 V. The gray-scale range displayed is 0.7 nm. Reprinted with permission from Ref. 83. Copyright 2011, American Institute of Physics.

top, center, and bottom plots show the measured normalized conductance from the InAs:Mn, InAs:Be, and InAs QDs respectively. For GaAs, the measured bandgap is slightly larger than the predicted 1.42 eV, likely due to TIBB. The effective bandgaps of the QDs are lower than those of GaAs, consistent with trends reported in the literature.[64,82,104,109,110]

In Fig. 9.14(a) and (b), spectra from the core and edge of the InAs QDs show well-defined effective band edges, without electronic states within the bandgap. In addition, the effective bandgap decreases laterally toward the QD core, presumably due to variations in [In], consistent with previous reports.[64,82,103,104] At the core of InAs:Mn QDs, shown in Fig. 9.14(a), the STS spectra are similar to those of the InAs QDs, with well-defined band edges and negligible electronic states within the bandgap. Near the edge of the InAs:Mn QDs, several additional spectral features, indicated by arrows in Fig. 9.14(b), are observed in the vicinity of E_e and E_h, most likely due to states associated with Mn. Thus, the electronic structure of the doped InAs:Mn QDs is altered most significantly near the QD edges, suggesting that the Mn dopants preferentially reside near the edges of the QD. This is further supported

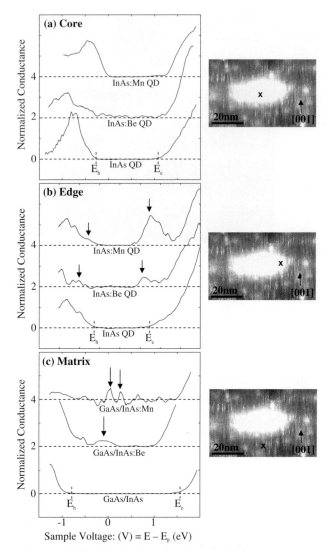

Fig. 9.14   Spatially-resolved STS spectra from the (a) core, (b) edge, and (c) GaAs matrix surrounding InAs, InAs:Be, and InAs:Mn QDs. STS spectra were collected at positions marked by an 'x' in the XSTM image to the right of the plots. For the STS plots, the sample voltage corresponds to the energy of the state relative to the Fermi level. The effective valence and conduction band edges are indicated by vertical dashed lines at negative and positive sample voltages, respectively. The spectra from the QD core are similar for all cases. In the spectra collected from the edge of the InAs:Mn QDs and from the GaAs near the edges of the InAs:Mn QDs, mid-gap features indicated by vertical arrows presumably correspond to Mn-induced electronic states. Reprinted with permission from Ref. 83. Copyright 2011, American Institute of Physics.

by the STS spectra from the GaAs matrix in the immediate vicinity of the QDs, shown in Fig. 9.14(c). For the GaAs matrix in the immediate vicinity of the InAs QDs, the STS spectrum shows well-defined band edges, without mid-gap spectral features. However, for the GaAs matrix in the immediate vicinity of the InAs:Mn QDs, several mid-gap features are observed. Similar trends were also observed for the InAs:Be QDs, as shown in Fig. 9.14.

9.6.2. LDOS calculations

We also computed spectra from the InAs:Mn QDs via LDOS calculated at the QD edge, QD core, and surrounding GaAs matrix, with and without Mn dopants at each location as shown in Fig. 9.15.[83,111,112] The LDOS calculations reveal InAs and InAs:Mn QD effective bandgaps that are lower than those of the surrounding GaAs matrix, consistent with the data discussed above. As shown in Fig. 9.15(a)–(c), the LDOS calculated at the QD core exhibit negligible electronic states within the gap. Thus, the QD core LDOS is predicted to be insensitive to the presence of Mn dopants, independent of the location of the Mn dopant within the QD or matrix.

We next consider the influence of Mn dopant location with respect to the QDs on the LDOS calculated at the QD edge (Fig. 9.15(d)–(f)) and in the surrounding GaAs matrix (Fig. 9.15(g)–(i)). For the case of the Mn dopants located at the QD core, negligible midgap spectral features are apparent in the LDOS, as shown in Fig. 9.15(d) and (g). However, when the Mn dopants are assumed to be located near the QD edge or in the surrounding GaAs matrix, additional spectral features are apparent in the LDOS, as indicated by arrows in Fig. 9.15(e), (f), and (h). Interestingly, for the "Mn at QD edge" case, the spectral features apparent in the LDOS calculated at the QD core (Fig. 9.15(b)), the QD edge (Fig. 9.15(e)), and in the surrounding GaAs matrix (Fig. 9.15(h)) are consistent with the STS data in Fig. 9.14. Thus, it is likely that the Mn dopants are preferentially located at the QD edges. Together, the STS data and the calculated LDOS reveal that the Mn dopants within the QD influence the electronic structure of both the QD and the surrounding GaAs matrix.

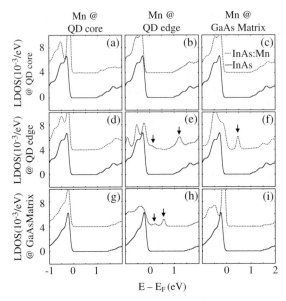

Fig. 9.15   The LDOS calculated at the QD core (a–c), QD edge (d–f), and GaAs matrix (g–i), using order (N) tight binding. LDOS from within and near the InAs:Mn (InAs) QDs are plotted as dashed (solid) lines towards the top (bottom). The LDOS was calculated assuming three different cases: Mn located at the QD core ((a), (d), (g)), Mn located at the QD edge ((b), (e), (h)), and Mn located in the surrounding GaAs matrix ((c), (f), (i)). The trends observed in the "Mn at QD edge" case are consistent with the trends observed in the measured STS spectra, suggesting that Mn dopants are preferentially located at the QD edges. Reprinted with permission from Ref. 83. Copyright 2011, American Institute of Physics.

This finding is in agreement with earlier studies of small, but not large, InAs:Mn QDs.[79] We therefore consider the nanoscale-size dependence of Mn incorporation in InAs QDs. Using a thermodynamic model described elsewhere,[105] we assume a spherical InAs nanoparticle with diameter 2 to 80 nm in a GaAs matrix. We calculate the Gibbs free energy assuming that Mn atoms at the QD core (surface) displace 4 In-As (3 In-As + 1 Ga-As) bonds. We use literature values for the In-As and Ga-As bond energies,[113] and a Mn-As bond energy calculated using the reported bulk solubility of Mn in GaAs.[114] The resulting Mn-induced change in Gibbs free energy, ΔG, as a function of QD diameter, is shown in Fig. 9.16, where the solid (dashed) lines represent the cases of a Mn

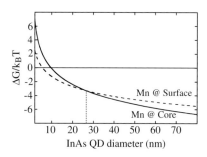

Fig. 9.16   Normalized Gibbs free energies for single Mn impurities, plotted as a function of InAs QD diameter. The solid and dashed curves represent the free energy for Mn impurities at the QD core and surface. Above (below) a critical QD diameter of 26 nm, core (surface) doping is expected to dominate. Reprinted with permission from Ref. 83. Copyright 2011, American Institute of Physics.

atom at the core (surface) of the InAs QD. When $\Delta G < 0$, Mn incorporation is predicted to occur. It is most energetically favorable for Mn to incorporate in the configuration that has the lowest value of ΔG. Thus, for the largest QDs, Mn is predicted to incorporate at the core, consistent with Ref. 78.[78] As the QD size is decreased to below 26 nm, Mn prefers to incorporate at the edges, as we show here.

9.7.   SUMMARY AND CONCLUSIONS

XSTM is a powerful method for resolving atomic features at interfaces in QD heterostructures because it allows direct observations of the spatial distribution of individual atoms on the surface. In addition, STS allows spatially-resolved electronic measurements within single layers of semiconductors, enabling the measurement of nanometer-scale variations in band gaps and band offsets across QDs. In this book chapter, we discussed the application of XSTM and STS to several important issues in self assembled III-V nanostructures, including (i) the influence of surrounding alloy layers on QD size, shape, density, and WL thickness, (ii) measurements of the electronic states of individual QDs as a function of position, including investigations into the influence of size, shape, and strain on the QD effective bandgap, and (iii) the influence of intentionally doping QDs on their electronic properties, and the relative locations of the dopants in the QD. A summary is provided below.

9.7.1. Influence of Alloy Layers on QD Size and Distribution

In Section 9.4, we presented an example XSTM study which focused on the influence of a variety of buffer and capping layers on the nanometer-scale structural properties of buried QDs and AQDs were examined at the nanometer-scale.[81] XSTM measurements revealed the diameters, heights, and WL thicknesses which are greater for the AQDs in comparison with the QDs. Furthermore, the lateral spacing between the AQDs was lower than the lateral spacing between the QDs, suggesting that the alloy buffer layer promotes an increase in dot density. It is likely that the alloy buffer and capping layers reduce the tendency for dot collapse, and that the diffusion of In from the alloy layers surrounding the WL lead to an increase in the apparent WL thickness. These results provide a valuable understanding of the relationship between the growth conditions and structure of semiconductor QDs, and are applicable to a wide range of similarly lattice-mismatched systems. In the future, it would be interesting to examine the influence of the varying alloy layer composition on the AQD size and distribution, and also to use a combination of plan-view and cross-sectional STM to separately examine the affects of alloy buffer and capping layers.

9.7.2. Quantum Dot Electronic States

In Section 9.5, we presented an example XSTM and STS study which focused on the nanoscale variations in the effective bandgap within InAs/GaAs QDs and the clustered regions of the WL.[82] The data revealed variations in the effective band gap across individual QDs both laterally and in the growth direction. Laterally, the effective bandgap decreased toward the QD core and vertically, the effective bandgap decreased in the growth direction. These results are consistent with an increase in [In] toward the center and top of the QD, suggesting that [In] variations dominated the variations in QD effective bandgap. Similarly, in the clustered regions of the WL, [In] variations dominated the variations in the WL effective bandgap. This work provides valuable insight into the influence of the real atomic structure of self-assembled QDs on the electronic properties. In the future, it would be interesting to examine the effective bandgap variations in both capped and uncapped QDs, using XSTM/STS in combination with plan-view STM/STS.

9.7.3. *Influence of Dopants on Quantum Dot Electronic States*

In this section, we presented an example XSTM and STS study in which the influence of dopants on the electronic states in the vicinity of individual nanostructures. Specifically, we analyzed the atomic-scale distribution of Mn dopants and their influence on the electronic states of InAs:Mn QDs and the surrounding GaAs matrix, using XSTM and STS, in comparison with order(N) tight-binding calculations of the LDOS.[83] This approach allowed us to deduce the positions of the dopants based on their influence on the LDOS. STS spectra from various locations in InAs and InAs:Mn QDs were compared with the LDOS of InAs and InAs:Mn QDs, calculated with Mn at the QD core, QD edge, or in the GaAs matrix. Spatially-resolved STS data and LDOS calculations revealed additional spectral features at the edges of InAs:Mn QDs and in the surrounding GaAs matrix, only when Mn atoms are assumed to be located near the QD edge. Therefore, it is likely that the Mn dopants preferentially reside near the edges of the QD. These results are consistent with the predictions of a thermodynamic model for the nanoscale-size dependence of dopant incorporation in nanostructures.[105] In the future, it would be interesting to examine the influence of dopants on the QD size and distribution, and to explore the structural and electronic effects of doping only the buffer or capping layer instead of doping the QDs.

REFERENCES

1. Y. Horikoshi, M. Kawashima, H. Yamaguchi, *Jpn. J. Appl. Phys.* **25**, L868 (1986).
2. Y. Horikoshi, *J. Cryst. Growth* **201–202**, 150 (1999).

402   *V. D. Dasika and R. S. Goldman*

3. J.-I. Chyi, T.-E. Nee, C.-T. Lee, J.-L. Shieh, J.-W. Pan, *J. Cryst. Growth* **175**, 777 (1997).
4. G. M. Guryanov, G. E. Cirlin, V. N. Petrov, N. K. Polyakov, A. O. Golubok, S. Y. Tipissev, V. B. Gubanov, Y. B. Samsonenko, N. N. Ledentsov, V. A. Shchukin, M. Grundmann, D. Bimberg, Z. I. Alferov, *Surf. Sci.* **352–354**, 651 (1996).
5. G. E. Cirlin, V. N. Petrov, V. G. Dubrovskii, A. O. Golubok, S. Y. Tipissev, G. M. Guryanov, M. V. Maximov, N. N. Ledentsov, D. Bimberg, *Czechoslovak J. Phys.* **47**, 379 (1997).
6. A. Bosacchi, P. Frigeri, S. Franchi, P. Allegri, V. Avanzini, *J. Cryst. Growth* **175**, 771 (1997).
7. W. Cheng, Z. Zhong, Y. Wu, Q. Huang, J. Zhou, *J. Cryst. Growth* **183**, 279 (1998).
8. J. Phillips, P. Bhattacharya, S. W. Kennerly, D. W. Beekman, M. Dutta, *IEEE J. Quant. Electron.* **35**, 936 (1999).
9. T. Yamauchi, Y. Ohyama, Y. Matsuba, M. Tabuchi, A. Nakamura, *Appl. Phys. Lett.* **79**, 2465 (2001).
10. T. Maltezopoulos, A. Bolz, C. Meyer, C. Heyn, W. Hansen, M. Morgenstern, R. Wiesendanger, *Phys. Rev. Lett.* **91**, 196804 (2003).
11. T. K. Johal, R. Rinaldi, A. Passaseo, R. Cingolani, A. Vasanelli, R. Ferreira, G. Bastard, *Phys. Rev. B* **66**, 075336 (2002).
12. O. Millo, D. Katz, Y. W. Cao, U. Banin, *Phys. Rev. B* **61**, 16773 (2000).
13. H. Y. Liu, M. Hopkinson, C. N. Harrison, M. J. Steer, R. Frith, I. R. Sellers, D. J. Mowbray, M. S. Skolnick, *J. Appl. Phys.* **93**, 2931 (2003).
14. H. Jiang, J. Singh, *Phys. Rev. B* **56**, 4696 (1997).
15. H. Shin, Y. H. Yoo, W. Lee, *J. Phys. D* **36**, 2612 (2003).
16. L. He, G. Bester, A. Zunger, *Phys. Rev. B* **70**, 235316 (2004).
17. H. Shin, E. Yoon, Y. H. Yoo, W. Lee, *J. Phys. Soc. Jpn.* **73**, 3378 (2004).
18. I. R. Sellers, H. Y. Liu, K. M. Groom, D. T. Childs, D. Robbins, T. J. Badcock, M. Hopkinson, D. J. Mowbray, M. S. Skolnick, *Electron. Lett.* **40**, 1412 (2004).
19. H. C. Yu, J. S. Wang, Y. K. Su, S. J. Chang, F. I. Lai, Y. H. Chang, H. C. Kuo, C. P. Sung, H. P. D. Yang, K. F. Lin, J. M. Wang, J. Y. Chi, R. S. Hsiao, S. Mikhrin, *IEEE Photonics Tech. Lett.* **18**, 418 (2006).
20. A. D. Stiff, S. Krishna, P. Bhattacharya, S. Kennerly, *Appl. Phys. Lett.* **79**, 421 (2001).
21. J. O. Winter, T. Y. Liu, B. A. Korgel, C. E. Schmidt, *Adv. Mater.* **13**, 1673 (2001).
22. G. Yusa, H. Sakaki, *Appl. Phys. Lett.* **70**, 345 (1997).
23. E. Biolatti, R. C. Iotti, P. Zanardi, F. Rossi, *Phys. Rev. Lett.* **85**, 5647 (2000).
24. D. Loss, D. P. DiVincenzo, *Phys. Rev. A* **57**, 120 (1998).
25. A. Ourmazd, F. H. Baumann, M. Bode, Y. Kim, *Ultramicroscopy* **34**, 237 (1990).
26. J. Stangl, V. Holy, G. Bauer, *Rev. Mod. Phys.* **76**, 725 (2004).
27. P. A. Crozier, M. Catalano, R. Cingolani, A. Passaseo, *Appl. Phys. Lett.* **79**, 3170 (2001).

28. P. Wang, A. L. Bleloch, M. Falke, P. J. Goodhew, J. Ng, M. Missous, *Appl. Phys. Lett.* **89**, 072111 (2006).
29. M. C. Joncour, M. N. Charasse, J. Burgeat, *J. Appl. Phys.* **58**, 3373 (1985).
30. T. Wiebach, M. Schmidbauer, M. Hanke, H. Raidt, R. Köhler, H. Wawra, *Phys. Rev. B* **61**, 5571 (2000).
31. T. F. Kelly, M. K. Miller, *Rev. Sci. Instrum.* **78**, 031101 (2007).
32. M. Muller, A. Cerezo, G. D. W. Smith, L. Chang, S. S. A. Gerstl, *Appl. Phys. Lett.* **92**, 233115 (2008).
33. A. D. Giddings, J. G. Keizer, M. Hara, G. J. Hamhuis, H. Yuasa, H. Fukuzawa, P. M. Koenraad, *Phys. Rev. B* **83**, 205308 (2011).
34. B. Gault, M. P. Moody, F. de Geuser, G. Tsafnat, A. La Fontaine, L. T. Stephenson, D. Haley, S. P. Ringer, *J. Appl. Phys.* **105**, 034913 (2009).
35. M. Arzberger, M. C. Amann, *Phys. Status Solidi B* **224**, 655 (2001).
36. V. M. Apalkov, T. Chakraborty, N. Ulbrich, D. Schuh, J. Bauer, G. Abstreiter, *Physica E* **24**, 272 (2004).
37. M. Bayer, A. Forchel, *Phys. Rev. B* **65**, 041308 (2002).
38. D. J. Mowbray, M. S. Skolnick, *J. Phys. D* **38**, 2059 (2005).
39. A. Kaneta, T. Izumi, K. Okamoto, Y. Kawakami, S. Fujita, Y. Narita, T. Inoue, T. Mukai, *Jpn. J. Appl. Phys.* **40**, 110 (2001).
40. T. H. Gfroerer, in *Encyclopedia of Analytical Chemistry* (John Wiley & Sons Ltd, 2000), p. 9209.
41. K. Brunner, U. Bockelmann, G. Abstreiter, M. Walther, G. Bohm, G. Trankle, G. Weimann, *Phys. Rev. Lett.* **69**, 3216 (1992).
42. J.-Y. Marzin, J.-M. Gérard, A. Izraël, D. Barrier, G. Bastard, *Phys. Rev. Lett.* **73**, 716 (1994).
43. T. W. Saucer, J.-E. Lee, A. J. Martin, D. Tien, J. M. Millunchick, V. Sih, *Solid State Commun.* **151**, 269 (2011).
44. D. Reuter, R. Roescu, M. Mehta, M. Richter, A. D. Wieck, *Physica E* **40**, 1961 (2008).
45. K. H. Schmidt, G. Medeiros-Ribeiro, M. Oestreich, P. M. Petroff, G. H. Döhler, *Phys. Rev. B* **54**, 11346 (1996).
46. P. N. Brounkov, A. Polimeni, S. T. Stoddart, M. Henini, L. Eaves, P. C. Main, A. R. Kovsh, Y. G. Musikhin, S. G. Konnikov, *Appl. Phys. Lett.* **73**, 1092 (1998).
47. G. Medeiros-Ribeiro, D. Leonard, P. M. Petroff, *Appl. Phys. Lett.* **66**, 1767 (1995).
48. P. Blood, *Semicond. Sci. Tech.* **1**, 7 (1986).
49. R. M. Feenstra, *Semiconduct. Sci. Technol.* **9**, 2157 (1994).
50. R. S. Goldman, *J. Phys. D* **37**, 163 (2004).
51. O. Albrektsen, Ph.D. Thesis, Technical University of Denmark, 1990.
52. R. J. Hamers, *Annu. Rev. Phys. Chem.* **40**, 531 (1989).
53. T. Tsuruoka, S. Ushioda, *J. Electron. Microsc.* **53**, 169 (2004).
54. B. Lita, Ph.D. Thesis, University of Michigan, 2001.
55. A. Stintz, G. T. Liu, A. L. Gray, R. Spillers, S. M. Delgado, K. J. Malloy, *J. Vac. Sci. Technol. B* **18**, 1496 (2000).

56. R. Songmuang, S. Kiravittaya, O. G. Schmidt, *J. Cryst. Growth* **249**, 416 (2003).
57. G. Costantini, A. Rastelli, C. Manzano, P. Acosta-Diaz, R. Songmuang, G. Katsaros, O. Schmidt, K. Kern, *Phys. Rev. Lett.* **96**, 226106 (2006).
58. J. M. Ulloa, C. Celebi, P. M. Koenraad, A. Simon, E. Gapihan, A. Letoublon, N. Bertru, I. Drouzas, D. J. Mowbray, M. J. Steer, M. Hopkinson, *J. Appl. Phys.* **101**, 081707 (2007).
59. T. Haga, M. Kataoka, N. Matsumura, S. Muto, Y. Nakata, N. Yokoyama, *Jpn. J. Appl. Phys.* **36**, L1113 (1997).
60. P. B. Joyce, T. J. Krzyzewski, G. R. Bell, B. A. Joyce, T. S. Jones, *Phys. Rev. B* **58**, R15981 (1998).
61. N. Liu, J. Tersoff, O. Baklenov, A. L. Holmes, Jr, C. K. Shih, *Phys. Rev. Lett.* **84**, 334 (2000).
62. A. Lenz, R. Timm, H. Eisele, C. Hennig, S. K. Becker, R. L. Sellin, U. W. Pohl, D. Bimberg, M. Dahne, *Appl. Phys. Lett.* **81**, 5150 (2002).
63. A. Urbieta, B. Grandidier, J. P. Nys, D. Deresmes, D. Stievenard, A. Lemaitre, G. Patriarche, Y. M. Niquet, *Phys. Rev. B* **77**, 155313 (2008).
64. D. M. Bruls, J. W. A. M. Vugs, P. M. Koenraad, M. S. Skolnick, M. Hopkinson, J. H. Wolter, *App. Phys. A* **72**, S205 (2001).
65. A. Lemaitre, G. Patriarche, F. Glas, *Appl. Phys. Lett.* **85**, 3717 (2004).
66. T. R. Ramachandran, A. Madhukar, I. Mukhametzhanov, R. Heitz, A. Kalburge, Q. Xie, P. Chen, *J. Vac. Sci. Technol. B* **16**, 1330 (1998).
67. B. Lita, R. S. Goldman, J. D. Phillips, P. K. Bhattacharya, *Appl. Phys. Lett.* **75**, 2797 (1999).
68. B. Lita, R. S. Goldman, J. D. Phillips, P. K. Bhattacharya, *Surf. Rev. Lett.* **7**, 539 (2000).
69. B. Shin, B. Lita, R. S. Goldman, J. D. Phillips, P. K. Bhattacharya, *Appl. Phys. Lett.* **81**, 1423 (2002).
70. P. Offermans, P. M. Koenraad, R. Notzel, J. H. Wolter, K. Pierz, *Appl. Phys. Lett.* **87**, 111903 (2005).
71. G. Sek, K. Ryczko, M. Motyka, J. Andrzejewski, K. Wysocka, J. Misiewicz, L. H. Li, A. Fiore, G. Patriarche, *J. Appl. Phys.* **101**, 63539 (2007).
72. H. Ohno, D. Chiba, F. Matsukura, T. Omiya, E. Abe, T. Dietl, Y. Ohno, K. Ohtani, *Nature* **408**, 944 (2000).
73. Y. Ohno, D. K. Young, B. Beschoten, F. Matsukura, H. Ohno, D. D. Awschalom, *Nature* **402**, 790 (1999).
74. S. Chakrabarti, M. A. Holub, P. Bhattacharya, T. D. Mishima, M. B. Santos, M. B. Johnson, D. A. Blom, *Nano Lett.* **5**, 209 (2004).
75. M. Holub, J. Shin, S. Chakrabarti, P. Bhattacharya, *Appl. Phys. Lett.* **87**, 091108 (2005).
76. M. Wang, R. P. Campion, A. W. Rushforth, K. W. Edmonds, C. T. Foxon, B. L. Gallagher, *Appl. Phys. Lett.* **93**, 132103 (2008).
77. A. Serres, G. Benassayag, M. Respaud, C. Armand, J. C. Pesant, A. Mari, Z. Liliental-Weber, A. Claverie, *Mater. Sci. Eng. B* **101**, 119 (2003).
78. M. Holub, S. Chakrabarti, S. Fathpour, P. Bhattacharya, Y. Lei, S. Ghosh, *Appl. Phys. Lett.* **85**, 973 (2004).

79. M. Bozkurt, V. A. Grant, J. M. Ulloa, R. P. Campion, C. T. Foxon, J. Marega, E. G. J. Salamo, P. M. Koenraad, *Appl. Phys. Lett.* **96**, 042108 (2010).
80. V. D. Dasika, Ph.D Thesis, University of Michigan, 2010.
81. V. D. Dasika, J. D. Song, W. J. Choi, N. K. Cho, J. I. Lee, R. S. Goldman, *Appl. Phys. Lett.* **95**, 163114 (2009).
82. V. D. Dasika, R. S. Goldman, J. D. Song, W. J. Choi, N. K. Cho, J. I. Lee, *J. Appl. Phys.* **106**, 014315 (2009).
83. V. D. Dasika, A. V. Semichaevsky, J. P. Petropoulos, J. C. Dibbern, A. M. Dangelewicz, M. Holub, P. K. Bhattacharya, J. M. O. Zide, H. T. Johnson, R. S. Goldman, *Appl. Phys. Lett.* **98**, 141907 (2011).
84. J. N. Gleason, M. E. Hjelmstad, V. D. Dasika, R. S. Goldman, S. Fathpour, S. Charkrabarti, P. K. Bhattacharya, *Appl. Phys. Lett.* **86**, 011911 (2005).
85. A. D. Schuyler, G. S. Chirikjian, J.-Q. Lu, H. T. Johnson, *Phys. Rev. E* **71**, 046701 (2005).
86. B. Shin, B. Lita, R. S. Goldman, J. D. Phillips, P. K. Bhattacharya, *Appl. Phys. Lett.* **81**, 1423 (2002).
87. This is an adaptation of a tunneling experiment in superconductivity.
88. C. K. Shih, R. M. Feenstra, G. V. Chandrashekhar, *Phys. Rev. B* **43**, 7913 (1991).
89. J. A. Stroscio, W. J. Kaiser, in *Scanning Tunneling Microscopy; Vol. 27* (Academic Press, Inc, 1993), p. 96.
90. R. M. Feenstra, *Phys. Rev. B* **50**, 4561 (1994).
91. R. M. Feenstra, J. A. Stroscio, *J. Vac. Sci. Technol. B* **5**, 923 (1987).
92. M. Weimer, J. Kramar, J. D. Baldeschwieler, *Phys. Rev. B* **39**, 5572 (1989).
93. S. Aloni, G. Haase, *J. Vac. Sci. Technol. B* **17**, 2651 (1999).
94. R. M. Feenstra, Y. Dong, M. P. Semtsiv, W. T. Masselink, *Nanotechnology* **18**, 044015 (2007).
95. Y. Dong, R. M. Feenstra, M. P. Semtsiv, W. T. Masselink, *J. Appl. Phys.* **103**, 073704 (2008).
96. S. Gwo, K.-J. Chao, C. K. Shih, K. Sadra, B. G. Streetman, *Phys. Rev. Lett.* **71**, 1883 (1993).
97. J. D. Song, Y. M. Park, J. C. Shin, J. G. Lim, Y. J. Park, W. J. Choi, I. K. Han, J. I. Lee, H. S. Kim, C. G. Park, *J. Appl. Phys.* **96**, 4122 (2004).
98. X.-D. Wang, N. Liu, C. K. Shih, S. Govindaraju, J. Holmes, A. L., *Appl. Phys. Lett.* **85**, 1356 (2004).
99. T. Walther, A. G. Cullis, D. J. Norris, M. Hopkinson, *Phys. Rev. Lett.* **86**, 2381 (2001).
100. H. R. Eisenberg, D. Kandel, *Phys. Rev. Lett.* **85**, 1286 (2000).
101. Y. Tu, J. Tersoff, *Phys. Rev. Lett.* **93**, 216101 (2004).
102. The measured bandgap is slightly larger than the predicted 1.42 eV, likely due to tip induced band bending.
103. J. Q. Lu, H. T. Johnson, V. D. Dasika, R. S. Goldman, *Appl. Phys. Lett.* **88**, 053109 (2006).
104. S. Gaan, G. He, R. M. Feenstra, J. Walker, E. Towe, *Appl. Phys. Lett.* **97**, 123110 (2010).

105. J. P. Petropoulos, T. R. Cristiani, P. B. Dongmo, J. M. O. Zide, *Nanotechnology* **22**, 245704 (2011).
106. J. C. Dibbern, V. D. Dasika, R. S. Goldman, (unpublished).
107. J. F. Chen, C. H. Chiang, Y. H. Wu, L. Chang, J. Y. Chi, *J. Appl. Phys.* **104**, 023509 (2008).
108. T. Matsuura, T. Miyamoto, T. Kageyama, M. Ohta, Y. Matsui, T. Furuhata, F. Koyama, *Jpn. J. Appl. Phys.* **43**, L605 (2004).
109. T. Yamauchi, Y. Matsuba, L. Bolotov, M. Tabuchi, A. Nakamura, *Appl. Phys. Lett.* **77**, 4368 (2000).
110. B. Legrand, B. Grandidier, J. P. Nys, D. Stievenard, J. M. Gerard, V. Thierry-Mieg, *Appl. Phys. Lett.* **73**, 96 (1998).
111. T. Dietl, H. Ohno, *MRS Bull.* **28**, 714 (2003).
112. J. Okabayashi, T. Mizokawa, D. D. Sarma, A. Fujimori, T. Slupinski, A. Oiwa, H. Munekata, *Phys. Rev. B J1 — PRB* **65**, 161203 (2002).
113. D. R. Lide, *CRC Handbook of Chemistry and Physics, 90th Edition* (CRC Press, Boca Raton, FL, 2010).
114. H. Ohno, A. Shen, F. Matsukara, A. Oiwa, A. Endo, S. Katsumoto, Y. Iye, *Appl. Phys. Lett.* **69**, 363 (1996).
115. K.-J. Chao, A. R. Smith, A. J. McDonald, D.-L. Kwong, B. G. Streetman, C.-K. Shih, **16**, 453 (1998).
116. J. N. Gleason, M. E. Hjelmstad, V. D. Dasika, R. S. Goldman, S. Fathpour, S. Charkrabarti, P. K. Bhattacharya, **86**, 011911 (2005).
117. S. Gwo, K.-J. Chao, C. K. Shih, **64**, 493 (1994).
118. B. Lita, R. S. Goldman, J. D. Phillips, P. K. Bhattacharya, **74**, 2824 (1999).
119. A. Vaterlaus, R. M. Feenstra, P. D. Kirchner, J. M. Woodall, G. D. Pettit, **11**, 1502 (1993).

10

ATOM PROBE TOMOGRAPHY FOR MICROELECTRONICS

David J. Larson, Ty J. Prosa, Dan Lawrence,
Brian P. Geiser, Clive M. Jones and Thomas F. Kelly

10.1. INTRODUCTION

The solid-state microelectronics industries are now 60 years old, and for most of that time there has been a need to understand the fine-scale makeup of devices both structurally and compositionally. Given the well-known trends toward smaller-scale devices,[1] the techniques that have been used to provide this information are required to continually improve their capabilities. As capabilities plateau, new techniques are developed or adapted to meet the need. The extant techniques at the time do not disappear typically but rather, the new techniques gain acceptance and take their place in parallel with established ones. Today we find ourselves at one of these junctures. With characteristic length scales in logic devices, memory devices, and data storage sensors well below 100 nm, there is a need for structural and compositional characterization at the sub-nanometer scale. The two standard bearers, transmission electron microscopy (TEM) (structural and compositional) and secondary-ion mass spectrometry (SIMS) (primarily compositional), can be extended to this length scale, but true three-dimensional (3D) characterization at the sub-nanometer scale is not currently within the grasp of either technique. There are no widely-adopted techniques capable of answering key characterization questions about nanoscale structures such as a buried interface with 1 nm roughness in a multilayer structure or dopant diffusion profiles in 3D surrounding a source/drain contact. This need for microscopy capabilities has created a characterization opportunity for atom probe tomography (APT).

APT is unique in the world of analytical instruments for its (1) spatial resolution in 3D (down to $\sim$0.2 nm), (2) analytical sensitivity ($<$10 atomic parts per million (appm)), and (3) high detection efficiency ($>$50%), combined with (4) its ability to detect all elements with equal efficiency and (5) its ability to detect all elements without need for *a priori* knowledge of the composition. All of these qualities make it a natural tool for the microelectronics industries.

During the past five years, a combination of developments in focused-ion-beam (FIB) technology and commercially available APT instruments has made it routine to apply APT to electronic structures. FIB instruments can now be used to selectively extract micron-sized pieces from metal alloys and semiconductor devices and shape them for APT.[2] The use of laser-pulsed field evaporation in APT has made it possible to analyze materials with low electrical conductivity including semiconductors, thin oxide films, and even bulk dielectrics.[3–10] Because the hardware to study real-world semiconducting materials has been available only recently, the applications are still evolving rapidly. For example, analysis of full CMOS transistors has been done by several groups (see the figures presented at the end of this chapter) but only after deprocessing the structure to remove some of the dielectrics surrounding the gate. There is great interest to analyze whole transistors without the need for deprocessing and this seems possible in the near future.

The hardware improvements in the early part of the 2000's which enabled larger sample volumes ($>$100 nm diameter by 500 nm long cylinders) analyzed at high data collection rates ($>10^6$ nm^3 per day)[11,12] have made the technique attractive to a much broader audience. New APT laboratories have been established around the world that cater to non-specialists. This expansion of the user-base has contributed to a

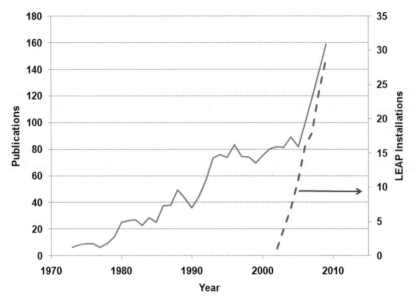

Fig. 10.1　Trends of atom-probe-related publications per year since 1974 (solid line-left axis). The numbers have been rising rapidly in the past five years as the number of commercial instruments (dashed line-right axis), such as LEAP® instruments, have increased.

recent upturn in the growth of scientific publications within the field over the past few years, Fig. 10.1.

APT's unique capability to identify and quantify individual chemical species in 3D is fast becoming a requirement for atomic-level microscopy. The technique is a natural complement to such other major microscopy techniques as TEM and SIMS with APT providing very high spatial resolution for chemical analysis. Well-equipped laboratories studying nanoscale phenomena will find APT has become an essential tool. The key to expanding the materials science problems accessible to APT has been, and will continue to be, the ability to craft specimens from a wide spectrum of materials in their as-available state. The development of lift-out techniques using a dual-beam FIB to extract site-specific speci-mens for analysis has greatly improved sampling productivity and made possible a wide range of new APT studies. This chapter describes the essence of these recent developments, with illustrations selected from microelectronics applications.

10.2. ATOM PROBE TOMOGRAPHY OVERVIEW

10.2.1. Origins

Atom probe tomography has its roots in field electron emission science and field ion microscopy developed principally by Müller from as early as 1935. The atom probe[13] was a major adaptation of the field ion microscope (FIM)[14] which was itself an adaptation of the field electron emission microscope.[15] The instrumentation has been undergoing radical changes throughout its history such that today's atom probes are millions of times more responsive to experimental characterization needs than their earliest predecessors. Reviews of these historical developments can be found elsewhere[16–21] including a recent review of the evolution of the instrumentation.[22] Most notable among these developments for microelectronic applications is the advent of commercially-available laser-pulsed atom probes[23] which make it possible to analyze semiconductor and electrically insulating materials following the pioneering work of Kellogg and Tsong.[24]

APT today produces 3D compositional images at the atomic scale with very high (<10 appm) analytical sensitivity. Because the fundamental data format is the 3D position and identity of individual atoms in a volume containing potentially hundreds of millions of atoms, many types of information may be gleaned. Elemental concentration may be determined in any subvolume size or shape simply by counting atoms. Concentration profiles may be obtained in any direction, even radially through a spherical feature or normal to any defined surface. Isoconcentration surfaces can be set to delineate and measure interfaces. Interatomic distribution functions can be determined for studying ordering, dopant interactions, cluster formation, and early stages of precipitation. Once the dataset is obtained, the quality and quantity of results is limited principally by the microscopist's ability to direct a computer to extract the information.

The unique power of atom probe microscopy lies in its ability to tie compositional information to structure. Other techniques can determine physical structure at sub-nanometer levels, but only APT has the capacity for broad compositional identification (and quantification) at these levels as well.

10.2.2.   Atom Probe Operation

The atom probe combines a time-of-flight (ToF) mass spectrometer
with a point-projection microscope capable of atomic-scale imaging.[13]
Combining this with position sensing, Fig. 10.2, enables the full 3D
reconstruction capabilities of APT. By applying a high voltage ($\sim$10 kV)
between the specimen and a counter electrode, a high electric field
($\sim10^{10}$ V/m) is created on the apex of a sharp ($<$100 nm radius) specimen held at cryogenic temperatures. Using a local-electrode,[11,25,26] as in
the local-electrode atom probe (LEAP®) of Fig. 10.2, allows the electric field to be applied selectively to a single specimen in an array of
microtips.[27] Atoms on the specimen apex are field evaporated as positive ions and accelerated toward the imaging detector. By pulsing the
evaporation, e.g. by using sub-nanosecond voltage pulses in addition to
the standing voltage, the flight time of each ion can be measured and
used to calculate the mass-to-charge-state ratio and thus its chemical
nature by ToF mass spectrometry. The $\sim$100 nm diameter area on the
apex is projected onto the roughly 100 mm diameter detector, giving a
magnification of 10^6. The original position of atoms on the specimen

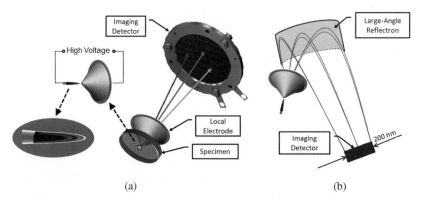

(a)   (b)

Fig. 10.2   (a) Schematic of a straight-flight-path atom probe. A specimen can be either a
single needle-shaped specimen (left) or a single microtip addressed from an array of microtip
specimens (right). In a local-electrode atom probe (LEAP), the specimen is opposed by a
local electrode which induces field evaporation of ions that fly out to an imaging detector.
This LEAP geometry is now common in commercial atom probes. (b) Schematic of a large-
angle reflectron atom probe. Here, ions fly out with a range of kinetic energies that are
corrected in ToF by variation in the flight path of the ions. Ions with large kinetic energies
travel longer flight paths than ions of the same mass-to-charge-state ratio giving both ions
the same ToF.

apex is determined from the hit position of the ions on the detector. Since atomic layers erode with regularity from the specimen surface, evaporation sequence number can be used to calculate an ion's position in z (direction parallel to the specimen long axis) in the original specimen with high precision. This combination of data allows 3D images of the elemental distributions to be reconstructed with near-atomic resolution. Further, atom probes detect all elements, isotopes, and higher-mass molecular ions.

10.2.3. Laser-Pulsed Atom Probe

Historically, voltage pulses have been used to generate pulsed field evaporation.[13] This requires that the specimen have high electrical conductivity $(>10^2\,S/cm)$, which limits the technique to metals and heavily doped semiconductors. Alternatively, short laser pulses $(<1\,ns)$ directed at the specimen apex can be used to induce a pulse in the evaporation rate for any material regardless of electrical conductivity.[24] Unlike the case of laser ionization mass analysis (LIMA), the temperature rises required for laser pulsed APT can be small: typically 100–200 K.[5,6] Data from Kellogg[5] show that for a given evaporation rate, the evaporation field decreases as temperature is increased. For proper quantitative operation, the voltage should be low enough to prevent evaporation between pulses. While voltage pulsing generates field evaporation at a constant temperature by raising the applied field, laser pulsing generates field evaporation at a constant field by raising the temperature. There have been suggestions that laser pulsing may invoke athermal field evaporation pulsing mechanisms[4] but experimental evidence has demonstrated that thermal pulsing dominates the process.[28–30]

Early experiments performed in the 1980's showed the promise of laser pulsing for studying semiconductor materials problems,[3,6,31,32] but the additional complexity in both instrument design and use prevented it from becoming mainstream. The advent of faster laser systems and well-engineered commercial atom probe instruments[23] has made laser-pulsed operation similar to that of voltage pulsing. Laser pulsing opens a broad universe of applications available for study by atom probe, such as silicon-based semiconductors and compound semiconductors,[33]

multilayer thin films,[34] ceramics[7,35–37] and even organics[38,39] (note that
the authors have made no attempt to be complete here in referencing
the various applications).

The challenge in developing a laser-pulsed atom probe was to create thermal pulses of sub-nanosecond duration at greater than 100 kHz
repetition rate that would work with poor thermal conductors. By
focusing the laser energy into a diffraction-limited spot of less than
3 μm diameter (4σ), a very small volume of the specimen is heated.
This makes it possible for the specimen to cool quickly and results
in an evaporation window (or time of departure spread) of less than
200 ps. It also makes it possible to analyze materials with poor thermal
diffusivity.

Another recent development is the migration to shorter wavelengths
in commercial atom probe systems. Early systems used either a laser fundamental wavelength of 800 nm (titanium/sapphire laser) or 1064 nm
(Nd/YAG laser), or a frequency-doubled green wavelength of 532 nm
(Nd/YAG laser). Frequency-tripled ultraviolet wavelengths of 355 nm[40]
or 343 nm[41] (Nd/YAG laser) were introduced commercially in 2009.[40–42]
One reason for utilizing the shorter wavelengths is that the absorption
depth (anomalous skin depth) decreases as the cube of the wavelength.
Not only can a smaller diffraction-limited spot be produced, but a shallower heated depth is also produced. Thus smaller heated volumes result
and more rapid cooling is produced. In all these cases, rapid cooling is
the key to achieving small time-of-departure spread which translates into
high mass resolving power.

Because it uses a relatively small, focused laser beam and a local
electrode, Cameca's LEAP 4000X (and its predecessor the LEAP 3000X)
is the only laser-pulsed atom probe with the capability to analyze
microtip arrays. The importance of this ability lies in the way specimens
may be configured. As shown below in Fig. 10.3, arrays of microtips may
be fabricated, for example, by micro electro mechanical system (MEMS)
processing methods for silicon. Such microtip arrays can serve many pur-
poses: (1) they may be analyzed directly after some form of processing
(thermal, implantation, etc.), (2) they may be coated with thin films
of inorganic or organic materials, and (3) they may serve as holders
for extractions from wafers or other bulk materials (including biological

Fig. 10.3 Scanning Electro-Microscope (SEM) image of a microtip array coupon. This array is prepared from a ⟨100⟩ silicon wafer. Individual microtips are ~100 microns tall and may be either sharp (~75 nm radius of curvature at apex) or flat topped (2 micron diameter flat).

materials). Material extraction methods (lift-out) make it possible to prepare specimens from a very small volume of material ($<1\,\mu m^3$) from a wide variety of sample types. As discussed below (see Fig. 10.10), small material wedge extractions may be mounted on a microtip and an entire array of microtips populated for analysis. Most of the applications described below were prepared via lift-out of material and subsequent transfer to microtip carrier arrays.

10.2.4. Instrumentation

Significant changes in atom probe instrumentation came about in the early part of the 2000's. Prior to this, the technique was constrained by a low speed of data collection (up to $\sim 10\,atoms/s \approx 10^6$ atoms/day) and a small field of view (about 15–20 nm diameter if a mass resolving power (MRP) of 500 was to be realized). With these limitations, the data collection rate translated to an analyzed depth of about 15–25 nm per day. Larger fields of view were possible, but only by sacrificing MRP. The challenge was to find a way to obtain high data collection rates and deliver high mass resolving power in a large-field-of-view configuration. This challenge has been met by at least two different approaches to date and these are illustrated in Fig. 10.2: the local-electrode atom probe in

Fig. 10.2(a), and the reflectron-compensated 3D atom probe (3DAP) in Fig. 10.2(b).

In a remote-electrode geometry, blunter tips require higher voltages to create the requisite electric field at the apex while projecting a larger area onto the same detector area. A local-electrode geometry[11,25,26] has the same trend, but for a given applied voltage the field is higher at the specimen and consequently the field of view is increased relative to the remote-electrode geometry. Practically speaking, the local-electrode typically doubles the field of view at a given voltage for commercial LEAP systems. As a result of faraday shielding of the flying ions from the time-varying voltage pulse, the potential energy gained by evaporated ions is also more uniform in LEAP which leads to improved mass resolution. The lower voltages also make it much easier to build high repetition rate voltage pulsers and achieve up to two orders of magnitude greater pulse repetition frequency (200 kHz) than previous atom probes (1 to 20 kHz). The sum of all these advantages produces an instrument that provides mass resolving power of 500 with a 200 nm field of view in aluminum at a pulse repetition rate of 250 kHz yielding 10^6 atoms/minute. This performance is achieved in a straight-flight-path geometry with a large detector that also delivers >55% detection efficiency for all ion types and which lends itself to even higher mass resolving power in laser pulsing and laser-pulse repetition rates up to 1 MHz.

Reflectron compensators[43] have been used in ToF atom probe systems for both one-dimensional instruments[44,45] and 3D instruments.[46,47] However, these instruments were limited in field of view by the chromatic aberrations of a flat reflectron. Panayi[48] developed the concept of a curved reflectron to overcome this limitation and introduced the device into a commercial 3DAP by Oxford nanoScience (ONS) in 2006. By 2007, a curved reflectron in a LEAP was commercially available as the LEAP 3000 HR. This system has the advantage that it yields high mass resolving power of greater than 1000 in a voltage-pulsed instrument with fields of view in excess of 150 nm diameter at full data collection rates. Furthermore, because of the greater flight path length of this geometry, the mass resolving power in laser-pulsed mode is also greater than 1000. The maximum laser pulsing rate is limited to 250 kHz due to the long flight times. The reflectron uses a fine mesh at the entrance surface to establish the entrance potential which has one major negative implication. Ions must pass twice through the mesh (in and out) and with 90% open-area meshes, this results in a geometric decrease in detection efficiency to about 80% of the best otherwise achieved.

10.2.5. Sensitivity

APT offers a unique combination of high analytical sensitivity coupled with high spatial resolution; however, there is often confusion regarding this fact in the microelectronics community, in part because of common expectations regarding elemental composition and sensitivity based on SIMS terminology.

SIMS is a very sensitive technique, but the analyzed sample volume required to achieve maximum sensitivity is not constrained, with analyzed areas often greater than several hundred square microns. Square microns of material cannot be analyzed by APT and so it can never compete with SIMS for sensitivity *at the micron scale*; however, the situation is very different at the nanoscale — the regime of individual device volumes.

Consider a volume of silicon doped to 2×10^{18} atoms/cm^3. In a $10 \times 10 \times 10$ nm^3 volume (containing $\sim$50 000 silicon atoms), the average total number of dopant atoms contained in that volume is *two*. If it were even possible for SIMS to measure a volume of this scale, ionization and detection efficiency limitations (only some 10^{-3}–10^{-6} of the atoms would be ionized and potentially detected) would allow measurement of only a few total atoms from this volume. Because APT has no ionization limitation and detection efficiency is limited only by current detector technology, $\sim$50% of all the atoms in the volume would be collected with a good statistical probability that at least 1 of the 2 dopant atoms would be detected. This particular combination of volume and doping level is right at the threshold of APT sensitivity (within the specified volume).

When analysis volume is a variable, discussing sensitivity in the "normally accepted" SIMS terminology of atoms/cm^3 is not appropriate because the analyzed volume is not a constant. A better way to express sensitivity in order to compare APT and SIMS is to use efficiency, which is independent of volume analyzed. Consequently in the nanoscale regime (i.e. when spatial information on a scale down to <10 nm is desired), when the analyzed volume is constrained, APT is the more sensitive and appropriate technique. Of course this discussion is somewhat simplified because factors other than ionization and detection efficiency can impact sensitivity (background signal in the mass spectrum for instance), but to first order, APT sensitivity is limited only by the number of ions contained in the small volume of interest with precision limited mainly by counting statistics.

10.3. SPATIAL RECONSTRUCTION IN ATOM PROBE TOMOGRAPHY

10.3.1. Overview

The main purpose of spatial reconstruction in APT is to convert the raw data collected by the detector into an accurate 3D spatial image of the atomic distribution of the specimen. For the purpose of this discussion we will neglect the specifics of the actual position-measurement hardware and just note that the raw data consists of a sequence of ion hit locations on the detector. Because of the atom-by-atom nature of APT data, at a high level the reconstruction problem consists of extracting 3D positions in specimen coordinates from a time series of 2D positions in detector coordinates. Composition analyses are typically performed by binning individually located atoms after the reconstruction process.

The standard reconstruction methodology[20,49-53] involves calculating a magnification in order to convert the detector hit positions (X and Y) into specimen-space coordinates in x and y (x and y are contained in the plane normal to the specimen axis at the current apex position). In the simplest convention, magnification is considered proportional to the distance between the sample center of curvature and the detector and inversely proportional to the product of the radius of curvature of the specimen, R_{spec}, and the angular magnification, α, which defines the point projection center and is usually treated as a single point for the entire field of view. It is generally accepted that the nominal value of the angular magnification is between one (radial projection) and 0.5 (stereographic projection) and its deviation from unity is due to the effects of the shank of the specimen compressing the electric field lines toward the detector (and away from a radial projection). A z coordinate is constructed by projecting to the surface of a geometrical model of the apex assuming the magnification computed above. To date, the model assumed has generally been limited to a spherical surface section. For each reconstructed atom, the z coordinate of the spherical model is adjusted by a small increment computed by assuming that the volume of the reconstructed atom was removed uniformly from the apex. The z increment is a function of R_{spec} and α, as well as a number of other instrumental and materials parameters,[50,51] such as detection efficiency and field-of-view.

Prior to the advent of wide-field-of-view instruments, the geometric assumptions described by Blavette,[49] and later applied by Bas,[50] were widely considered the standard global reconstruction technique. This model assumes that the original shape of acquired volumes is small enough in lateral extent to be considered cylindrical and the radius of curvature of the specimen is determined atom-by-atom by the specimen voltage. In the early reconstructions, the actual shank angle was ignored and assumed to be zero in the calculation of the volume increment. Walck published a description of specimen geometry that could be used to compute depth increments that are large field of view and included shank-angle effects.[53,54]

Although a complete review of the current state of APT reconstruction is beyond the scope of the current article, some relevant references are provided for the interested reader.[51–62]

10.3.2. Spatial Resolution

Spatial resolution is a local measurement. That is, it should express the ultimate performance of the instrument and not be influenced by other sources of imperfection such as specimen-related voltage instabilities, imperfect reconstruction procedures or trajectory aberrations near a low-index zone axis or phase boundary. Thus, like most microscopies, the spatial resolution achieved in APT depends on the specimen as well as the instrument. There are imperfections that degrade the achievable resolution due to:

(a) the specimen temperature,
(b) detector and optical limitations,
(c) reconstruction errors,
(d) trajectory aberrations in monophase materials, and
(e) local magnification in polyphase materials.

Thus again, like most microscopies, the spatial resolution achieved in a given dataset will vary within the image and may never reach the ultimate potential of the technique. Nonetheless, it is instructive to consider the best demonstrated spatial resolution as a starting point and then consider the imperfections individually.

Vurpillot et al.[63] defined spatial resolution in APT as the inverse of the width of the damping function that is superimposed on the reciprocal space structure of an atom probe dataset. This parameter gives information about the uncertainty in interatomic distances as determined by APT. Geiser et al.[64] introduced the spatial distribution map (SDM) as a tool to average real space information about interatomic distances in an APT dataset and measured interatomic separation variations in the data (functionally equivalent to the work of Vurpillot et al.[63]). Kelly et al.[65] summarized these approaches and proposed a standard measurement of spatial resolution using tungsten specimens. Gault et al. proposed an expression for resolution in crystalline systems based on statistical considerations of planar spacing and temperature.[66,67]

These definitions all pursue the same essential information: the limits of the spatial resolution as revealed by the real-space atom-location uncertainty/reciprocal-space damping function. Furthermore, there are two physically different mechanisms involved with resolving information in the longitudinal and lateral directions of APT, respectively, and measurements for both are required. In each of these papers referenced above,[63–67] a lateral resolution of about 0.15 nm has been demonstrated in tungsten at ~50 K. Similarly, the longitudinal (depth) spatial resolution has consistently been found to be about 0.06 nm.

10.3.3. Imperfections in Spatial Positioning

As mentioned above, there are many factors that can cause the observed spatial resolution to be less than optimal. They will be discussed individually below.

(a) *Effect of Temperature* The achievable resolution in both FIM and APT is a function of specimen temperature.[16,17,66–68] Lateral thermal motion and surface diffusion can both contribute to a blurring of determined atom positions. In order to obtain high spatial resolution, specimens should typically remain below a base homologous temperature of approximately 0.1. Even with tungsten specimens, the image resolution will degrade discernibly between 30 K and 50 K. It is therefore common practice to operate atom probes with base specimen temperatures of 20 to 80 K depending on the material. This is common practice for both voltage and laser pulsing. However, with laser pulsing, the specimen apex temperature is necessarily being raised a large amount (about 5× base temperature) for a short time (∼100–500 ps). Thus the best spatial resolution is generally achieved with voltage pulsing. It is possible to reach temperatures with laser pulsing where it is readily apparent that the spatial resolution of the image has been severely compromised. No level of improvement in the limitations listed below will compensate for thermally degraded spatial resolution.

(b) *Detector and Optical Limitations* The position-sensitive detector used in an atom probe must have sufficient pixel count to faithfully record the achievable resolution. For example, in a 100 nm diameter field of view, there will be about 500 atoms on the diameter of the image. The detector will need at least 1000 pixels across the diameter to position atoms with precision. Stated differently, 1000 pixels in 100 nm means that 0.1 nm spatial resolution is the geometric limitation of the detector. As specimens get blunter during an analysis, the magnification decreases and a greater number of atoms are imaged on the constant detector diameter. It is therefore possible that in the early part of an analysis the optimal spatial resolution is achieved but in the latter stages of an analysis, the spatial resolution is limited by detector pixel size.

Optically, the spatial resolution can be limited by voltage instabilities that cause variations in magnification. These aberrations affect
only the lateral resolution. In instruments, including reflectron energy
compensators,[43,45–48] electrostatic surfaces are employed to provide
improved mass resolution. In some designs, ion trajectories may pass
multiple times through a metallic mesh. The electric field close to the
wires can be anisotropic and cause noticeable trajectory distortions.

(c) *Basic Reconstruction Algorithm Limitations* The quality of reconstruction algorithms is a central question in spatial resolution
achieved in APT. A thorough treatment of this topic is beyond
the scope of this chapter; however there are some key elements that
should be discussed. As described above, the standard reconstruction algorithms today have been adapted to large-acceptance-angle
images in commercial software.[53] To the extent that the approxima-
tions made are accurate, there is often uncertainty as to the parameter values to be used. For example, in the absence of microscopy
data from the initial end form, the starting specimen radius, R_{spec},
is inferred from the evaporation field, F, of the specimen material
and the specimen voltage, V, required to sustain field evaporation
at a specified rate,[69] $R_{spec} = V/kF$, where k is a geometric field-
enhancement factor. The evaporation field is typically a function of
temperature[70] and sensitive to specimen composition[71] and crystalline orientation[72] while k will depend on the details of the specimen shape[52,69,73] and spatial position relative to other pieces of
the flight-path.[74–76] Procedures for determining global values for
these parameters from crystalline specimen APT data have been
suggested by Gault *et al.*[51]

In addition, current algorithms implemented still assume a spherical apex on the specimen that is tangential to a cone at the base of
the spherical cap. Many investigators have observed that the spherical shape is incorrect in both experimental and simulated images
of field-evaporated specimens.[57–59,61,77–82] Real specimens have a
blunter apex near the center with greater curvature near the edges,
Fig. 10.4. The evaporated cap also is not tangential to the cone of
the specimen. The principle effect of this is a non-uniform angular
magnification[78] with distance from the central axis, i.e. $\alpha = \alpha(R)$

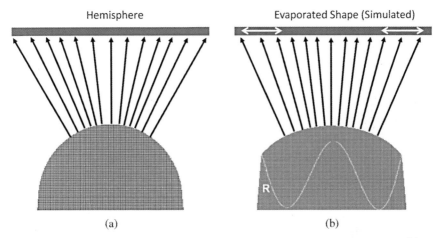

Hemisphere   Evaporated Shape (Simulated)

(a)   (b)

Fig. 10.4   Simulations of field evaporation from a crystalline specimen end form. (a) A spherical-cap end form with tangential continuity at the base starting out evolves into (b) a non-spherical shape with a discontinuity in the surface where it meets the base. Instead of a uniform curvature across the apex, the curvature varies with radial position from the center as shown by the white sinusoidal line on the right image. The effect of a non-spherical shape is an anisotropic magnification across the surface, even in the case of simple single-phase structures. This angular magnification must evolve with depth if the base is a cone of non-zero angle.

where $R = x^2 + y^2$ is the radial distance on the detector away from the center. The lateral arrows in Fig. 10.4(b) indicate regions of increased magnification due to increased apex curvature. Furthermore, when the specimen shank has some taper, this function varies with depth in an image, z, i.e. $\alpha = \alpha(r, z)$ where r is the radial position in the image. This effect is being introduced into commercial reconstruction algorithms. Without this correction, planar features in data will not be reconstructed as planar.[59] All materials including amorphous materials are expected to need this angular magnification function.

Although the following two aberrations are arguably reconstruction limitations, they are specimen dependent and have therefore been separated out for clarity.

(d) *Monophase Aberrations* The next level of complexity that limits spatial resolution is an aberration that is found only in crystalline materials. Atoms generally field evaporate from the edges of closely packed planes and not from the interior of a plane.[16] As a result, in

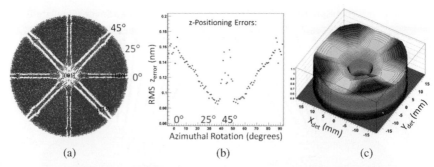

Fig. 10.5   (a) Detector hit histogram of a simulation of field evaporation from a face-centered cubic specimen with the $\langle 100 \rangle$ centered on the tip axis. Note the region of low hit density around the $\langle 100 \rangle$ facet. Not only are there errors in lateral positioning of atoms caused by the faceting, but the z position also accrues errors as shown in (b). (c) is an automatically optimized angular-magnification function for this case. This angular magnification must evolve with depth if the base is a cone of non-zero angle. Figures taken from Ref. 59.

areas of high atomic packing density, typically normal to low-index crystal directions, planar facets form on the evaporating specimen. The lateral extent of these planes depends on their level of packing. These planar facets create deviations from a smooth surface and result in local non-uniformities in angular magnification, i.e. $\alpha = \alpha\,(r, z, \varphi)$ where φ is the azimuthal angle about a central axis of the specimen. This effect is illustrated in Fig. 10.5 from the detector hit histogram for a simulated face-centered cubic material oriented with $\langle 100 \rangle$ centered on the tip axis. The details of faceting in the simulation replicate experimental results quite well. A key advantage of simulations for studying reconstruction algorithms is that the starting positions of all atoms are known. With this information, it is possible to determine the errors introduced by the algorithm. Not only must the lateral (x, y) position error be treated, but in Fig. 10.5(b) the longitudinal (z) position errors are shown as a function of azimuthal angle. From these data, the function needed to correct for positioning errors may be computed[59] as shown in Fig. 10.5(c) for a small z increment. In practice, this type of function must be determined as a function of z for arbitrary orientations and specific structures. Different alloys of the same crystal structure may also have different field evaporation behavior and unique functions may be needed for each alloy.

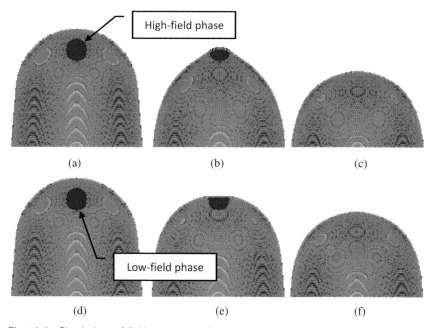

Fig. 10.6   Simulations of field evaporation from a crystalline specimen apex which contains a spherical second phase that requires a different evaporation field from the surrounding material. (a)–(c) Represent the situation where the spherical second phase has a higher evaporation field than the surrounding material. (a) Shows the specimen after field evaporation of the base material. (b) and (c) Show sequential snapshots of this simulation after some amount of field evaporation has occurred. (d)–(f) Represent the situation where the spherical second phase has a lower evaporation field than the surrounding material. The distortion of the end form that results when this phase emerges at the surface is evident in (b) and (e). The distortion creates trajectory aberrations (local magnification differences) for both phases near the interface and may create mixing of those atoms with matrix atoms. All simulations shown in this figure are calculated using Ref. 85.

(e) *Polyphase Aberrations* Specimens with multiple phases present the most challenging reconstruction opportunities. In principle, not only do we need to take all of the above effects into account, but they must be done for phases within phases. Consider the case of a spherical second phase as shown in Fig. 10.6. After some evaporation, the specimen will form an endcap as in Fig. 10(b) above, Fig. 10.6(a). Once the second phase emerges at the surface, Fig. 10.6(b), the entire endcap is altered and trajectories of atoms in both phases are different from a monophase material. In general, the atoms of a high-field phase will see greater angular magnification than the low-field phase.[80,83,84] In the limits of large or small relative size of the second phase and small difference in evaporation field, this effect disappears. If the evaporation field of the second phase is less than the surrounding material, Figs. 10.6(d)–(f), the situation is reversed.

In either case, the effects of the resulting shapes on ion trajectories from either phase raises the possibility of ion crossings. If ion paths from the two phases do not cross, then this aberration should be resolvable in principle. If ion paths do cross, the task of correcting this aberration gets more challenging and may make this aberration uncorrectable in a traditional atom-by-atom treatment.

10.3.4.   Development Directions for APT Reconstruction

Correcting the aberrations mentioned above is the subject of current research in the field of APT. Indeed, the specimen *is* the primary optic of the atom probe microscope; hence, all of the models above concerning reconstruction require knowledge of appropriate geometric parameters for the algorithms. Since specimens can vary markedly, the fact that we often lack this crucial information regarding the specimen geometry to set these parameters is a limitation: a situation referred to as unconstrained reconstruction. If we admit physical information such as the shape of second phases to constrain the reconstruction, then the prospects for correction improve dramatically. For example, electron microscope images of the specimen may be used to constrain the reconstruction process in APT.[57,82,86,87] This multi-modal approach to atomic-scale microscopy may be necessary ultimately to perfect reconstruction in an atom probe as described further below.[88]

10.3.4.1.   The role of simulations

In the absence of simultaneous multi-modal imaging, other methods must be used to infer the evolving shape of the apex and the resulting variations in magnification. As alluded to above, simulation of various aspects of the processes involved in APT has been of interest for many years.[58,59,85,89–97] Because of the critical nature of the shape of the specimen[58] in the data-reconstruction process described above, recent work has focused on simulating evaporated shapes,[58,59,61,96,97] confirming the accuracy of these shapes using electron microscopy,[57,58,81] and using such information in the reconstruction process.[59-62]

Simulation engines for the field evaporation process in APT have been developed by several groups in the past decade.[85,89,98] To date, these simulations are very simple in that evaporation fields are considered globally constant for a given species. The evaporation mechanisms are of two basic types: (1) the evaporation order is determined primarily from local coordination number and electrostatic effects are neglected and (2) the evaporation order is determined purely from electrostatics and detailed binding or surface effects are ignored. Surface diffusion,[99] dynamic thermal effects,[82] and configuration-dependent binding are still a challenge for these tools.

At least qualitatively, the simulations have improved to the point that they are capable of providing useful diagnostic information in systems of current interest.[58,59,61,84] Evaporation simulations are also currently playing an important role in validating reconstruction algorithms since they allow some level of artifact quantification. An example of this is the effect of a post-projection z-correction illustrated on synthetic and real data by Vurpillot *et al.*[97] As indicated above, studies are underway, investigating the role of non-uniformity in angular-magnification functions. As another illustration, Fig. 10.7 shows an example of how the simulation of the evolving shape of a specimen could be used to calculate the predicted evolution in the local radius for use in reconstruction. As field evaporation proceeds through thin layers of varying evaporation fields,[61] Figs. 10.7(a)–(b), the radius at the center of the specimen varies accordingly, as shown in Fig. 10.7(c).

Although it is clear these simulation tools will play an increasingly important role in the reconstruction of APT data, especially in the reconstruction of structured materials, their use is still very challenging. Because the evaporation process is intrinsically serial, it is not obvious how parallel computing can help speed the process and computational performance can be an important consideration. More importantly, it is expected that accurate quantitative matching of APT images with synthetic data will require further development in the physics of the simulation tools.

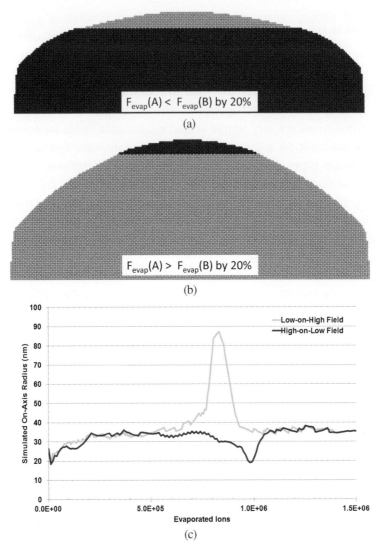

Fig. 10.7 (a) Evaporated specimen shape for a specimen containing a low-evaporation-field layer above a high-evaporation-field material ($1.2 \times F_A = F_B$) illustrating the blunting effect caused by different layers having different evaporation fields. (b) Simulated shape for a specimen containing a high-evapo field layer above a low-evaporation-field material. (c) Estimation of the specimen radius (on the axis) during the simulated evaporation shown in (a) and (b). Figures in (a) and (b) are taken from Ref. 59.

10.3.4.2. Longer-term outlook

Ideally, reconstruction algorithms would incorporate information from both multi-modal microscopy and simulation, as shown schematically in

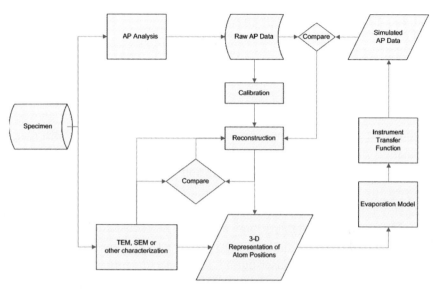

Fig. 10.8   A schematic of information flow in an iterative reconstruction procedure incorporating information from multi-modal microscopy and simulation.

Fig. 10.8. In this respect, imaging in an atom probe is similar to phase contrast imaging in high resolution TEM. We would like ideally to have a closed information loop. That is, if we place a specimen in a microscope and record an image with that microscope, then we should be able to compute or infer the unique specimen that produced the image. Unfortunately, neither of these techniques is closed loop, and we must record the image and then guess at the specimen shape. The image that the microscope would produce from that specimen is then computed from an estimated image transfer function, and the actual image and computed image are compared. This is an open loop process but it most often works. In order to have this open loop process for APT, we must be able to compute the image from a presumed specimen. In fact, it has been noted by Miller and Kelly[88] that the information loop from TEM/STEM and that from APT are complementary. By combining these two sources of information, a closed-loop image formation process is obtained. This concept is an essential element of a proposal to build an atom probe tomograph within a high-resolution scanning TEM. This project is titled "The ATOM Project," and is being considered for funding by the US Department of Energy.

10.4. SPECIMEN PREPARATION

10.4.1. Overview

The essential geometry of an atom probe specimen today is a sharp needle that tapers with a modest angle ($\sim10°$ semi-angle) to an apex radius that is 50 to 200 nm. The small specimen size provides both magnification of surface features and creation of the high (~10–50 V/nm) electric field necessary for field evaporation to occur. Because the needle-shaped specimen is the primary optic of the microscope, specimen preparation is critical to successful APT analysis.

The general availability and adoption of focused-ion-beam (FIB) specimen preparation methods over the past ten years, coupled with commercially available laser-pulsed APT instruments, has ushered in a period of rapid acceleration of APT application to new materials, especially microelectronic devices. The ability to extract a "nano-biopsy" of a specific component within the real microelectronics device, and place it at the specimen apex, i.e. 3D site-specific lift out, has been key in enabling the applications summarized later in this article.

The earliest FIB-based methods to fabricate specimens for APT relied on attaching a volume of material to the end of a needle (using non-FIB methods) and shaping the end-form for APT analysis with FIB.[100–103] Optimal annular milling methods were developed that provided for improved emitter shapes,[104–108] while adoption of TEM lift-out methods[2,108–110] that could utilize microtip array carrier tips[27,111] improved repeatability and reduced preparation time. Today, many applications and variations of the standard lift-out and sharpening methods have been reported.[108,112–115] These include variations enabling analysis parallel to the original specimen surface ("cross-section" orientation)[116] or inverted relative to the original specimen surface ("backside" orientation).[87,117]

For microelectronic devices, FIB-based site-specific specimen preparation is required. The general goal of such preparation is to place a region of interest (ROI) in the apex of a needle with the geometry described previously. The material should be pristine (free of FIB-implantation artifacts[109]) and as near the apex as is practical. Because the field-evaporation process applies a high stress to the specimen apex region, many specimens are prone to failure during analysis due to structural defects or undesirable material interfacial properties (high and low evaporation field material interfaces, as shown in Fig. 10.7, or interfaces with poor adhesion). Therefore production of multiple, nearly identical, APT specimens is desired to test analysis repeatability and improve the chances for successful analyses.

Some preparatory preprocessing or deprocessing of the device structure may be necessary before lift-out and sharpening can proceed. Preprocessing might include applying a protective sacrificial cap to the top surface. Deprocessing might include removal of unnecessary microelectronic superstructure material to allow easier access to the ROI. This may be as simple as mechanically scraping away epoxy or other protective coverings or more sophisticated physical (FIB milling) or chemical (etching) methods for removing superstructure layers that are not of interest. Each of these methods is generally described below.

Once the ROI is properly protected and positioned with respect to the top surface of the host material, the lift-out process can be initiated. A multi-step process is described below where material is removed during a milling procedure to produce a cantilevered wedge of material containing the ROI. The wedge is then removed by a micromanipulator and transferred to carrier tips. Subsequent sharpening of the transferred material via an annular milling procedure followed by a low energy "clean-up" step is also described. Variations to the standard lift-out process are discussed that allow for analyses to proceed in directions other than the surface normal. Finally, preparation of specimens for hybrid TEM/APT analysis is described.

10.4.2. Capping Methods

In the context of APT, capping refers to the application of a sacrificial layer to the top surface of a sample. This procedure both protects the underlying material from the 30 kV gallium ions used during the lift-out process and subsequent tip shaping, and also provides a layer with repeatable ion milling properties to enable consistent and predictable tip shapes. Typical Ga ion implantation due to FIB milling is illustrated in

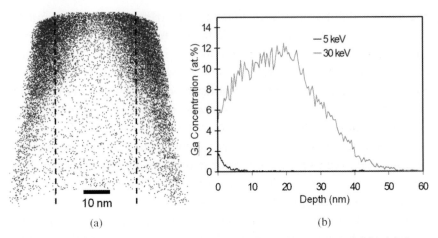

(a)   (b)

Fig. 10.9   Typical gallium implantation caused by FIB-milling at 5 and 30 keV. (a) Atom map obtained from APT of a 5 keV shaped specimen. The dark dots represent individual gallium atoms. The dashed lines represent the analysis volume used to produce the atomic gallium composition as a function of depth shown in (b). (b) A comparison of the gallium composition profile with depth for specimens shaped with 5 and 30 keV Ga ions.

Fig. 10.9, which is from a doped silicon wafer prepared using standard lift-out methods and annular milling techniques (for details see next sections). Final milling was performed with 30 and 5 keV Ga ions to investigate implantation depths.

Figure 10.9(a) shows an atom map from an analyzed tip that has been shaped with 5 keV Ga ions. It is important to note that the analyzed volume of the tip is only the central portion (dashed lines) of the real tip-shaped specimen. As the analysis proceeds, the Ga implantation is greatly reduced as the analysis becomes farther removed from the original tip surface: first laterally, then in depth. Figure 10.9(b) compares the penetration depth of 30 and 5 keV ions. The implantation region is clearly reduced to ~5 to 10 nm for 5 keV. For the case of final specimen shaping using 2 keV ions, almost no residual Ga ions were detected.[109,118]

Because APT removes material as the analysis proceeds, damaged regions of the material can be removed prior to data collection and only undamaged volumes are typically retained for detailed analysis. Because many materials are especially prone to failure during application of a high electric field, the typical volume of material provided by an APT measurement may not be sufficient to allow analysis beyond the damage region. In addition, the large damage regions themselves may (a) be the cause for premature failure of the analysis, (b) lead to intermixing of phases and regions of different compositions, or (c) turn crystalline regions amorphous.[102,108] It is therefore generally considered best practice to remove as much of the damaged region of the specimen as is practical.

For specimens where the original surface material is part of the analysis ROI, or when surface properties promote increased damage depths (presence of grains or interfaces running into the surface of the material promoting ion channeling), it is desirable to retain some of the capping material in the final tip for analysis to ensure damage-free analysis volumes. When the material surface/cap interface is part of the analysis, strong interfacial adhesion is critical for specimen survival. The capping material needs to be consistent with good adhesion[119] and good tip shaping (ion milling) properties, and possess good field evaporation properties. Commonly used capping materials for microelectronics applications include nickel, chromium, and polysilicon. Multiple capping materials can also be used to manufacture a multilayer cap with a combination of desired properties — nickel or chromium for good adhesion, gold or platinum for a highly visible end-pointing layer (with secondary electron imaging), or platinum for durability under ion milling, etc.

10.4.3. Standard Lift-Out Process

For material structures where (a) the ROI can be deduced from surface features, (b) the ROI resides near the surface, and (c) the analysis direction is intended to proceed from the top surface down into the material (this orientation will be termed "top-down" in the current work), the standard lift-out preparation is appropriate.[108,109] A number of variations of the general method are known to exist, of which only some have been detailed in the literature. The specific steps illustrated below describe the method most commonly used by the current authors.

Figure 10.10 illustrates the general steps followed during a standard lift-out procedure. First, an optional protective capping layer may be applied over the entire specimen surface. For the illustrated example, no capping layer is necessary because the ROI is sufficiently below the top

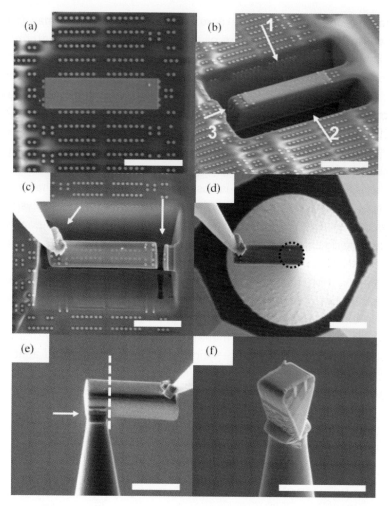

Fig. 10.10 The steps for performing site-specific 3D lift out with the standard lift-out procedure. (a) The area containing the region of interest (ROI) is identified and a FIB-deposited protective strip (Pt) is placed over the region. (b) Material is removed around three sides of the region (arrows) as well as underneath to produce a long cantilevered wedge of material containing the ROI. (c) The wedge is removed by first attaching a micromanipulator to one end of the wedge (left arrow) and then cutting the wedge free from the substrate (right arrow). (d) The wedge is next transferred to a carrier microtip (black dashed circle indicates location of flat $2\,\mu$m tip). (e) After attaching the wedge to the carrier tip surface with FIB-deposited Pt (arrow), the wedge is cut free of the carrier tip (dashed line) for transfer to additional tips. (f) The final mounted wedge-section is shown with FIB-deposited Pt at each wedge-tip interface. Note all scale bars are $5\,\mu$m.

surface. Next, a FIB-deposited platinum strip is added to protect the surface and to demarcate the region for extraction (Fig. 10.10(a)). The Pt layer is typically 2–3 μm wide and $\sim$100 nm thick, while the length of the Pt layer depends on the geometry of the ROI and the number of desired specimens to be made from the extraction. In this example, the strip spans the length of a continuous region containing identical ROIs ($\sim$10 μm).

Material is next removed around the ROI to create a cantilevered mass of material for lift-out (Fig. 10.10(b)). In this case, trench 1 (see arrow) is created by tilting the stage to 22° and milling a ($\sim$2 $\times$ 10) μm^2 rectangular pattern with 6 nA ion current. The ion beam is scanned parallel to the long axis of the trench starting far from the Pt strip and proceeding to the near edge. Trench 2 is created using the same procedure after rotating the stage 180°. Milling is completed when trenches 1 and 2 meet beneath the ROI. Trench 3 is then cut using a $\sim$1–2 μm long rectangular pattern of sufficient width to cut across the entire wedge leaving behind a cantilevered mass of material as shown.

After returning the FIB stage to 0° stage tilt, a micromanipulator is lowered into contact with the free end of the cantilever (Fig. 10.10(c)). Sufficient FIB-deposited Pt is used to secure the micromanipulator to the top surface of the wedge (left arrow). Once the manipulator is secured, the wedge is cut free of the bulk by cutting a $\sim$1–2 μm long rectangular pattern again with width sufficient to cut across the entire wedge leaving behind a cantilever of material (right arrow).

Once the wedge and manipulator are free of the bulk, the wedge can be transferred to a carrier tip (Fig. 10.10(d)). In this image, the wedge is placed above a $\sim$2 μm diameter, flat, silicon microtip post (the post is centered within the black dashed circle). The microtip posts are manufactured on a 450 μm array with each post having an overall height of $\sim$100 μm above a planar surface. This height and spacing combination is sufficient to allow for rapid transfer of wedge material to multiple posts and to allow independent analysis of each post without field-evaporation of neighboring specimens. The wedge is carefully lowered until it comes into direct contact with the top surface of the post. A ($\sim$1 $\times$ 1 $\times$ 0.5) μm^3 Pt patch is deposited over the wedge-post interface to secure the wedge to the post (Fig. 10.10(e) arrow). The wedge is then cut from the post as indicated by the dashed line in Fig. 10.10(e), and the wedge is propagated to additional posts until the wedge is completely consumed.

Once propagation of the wedge is finished, the stage is rotated 180° so that a second Pt patch can be applied to the opposite wedge-post interface of each mounted post. The final mounted wedge slice is shown in Fig. 10.10(f) where the Pt deposits are clearly visible on each side of the wedge-post interface.

10.4.4.   Sharpening Process

The specimen dimensions required for APT analysis are material and instrument dependent. The user must consider this fact, as well as the voltage-range limitation at which an atom probe can operate, when preparing specimens.

Features of interest which are relatively large need to be positioned below the initial apex of a specimen. A general rule of thumb is to prepare a specimen which has a lateral diameter at the ROI depth position of about 2X the lateral dimension of the ROI.

The process of converting a wedge extraction into a sub-200 nm diameter sharp needle is accomplished through a series of annular milling[105] steps followed by a low-energy FIB clean-up step,[109,118] Fig. 10.11. Illustrated in Fig. 10.11(a) is a transistor centered within the wedge shown in Fig. 10.10. Some key components of the transistor are visible in the inset and surround the polysilicon gate (dashed box). Visible to the right and left of the gate are the spacer regions, while to the top and bottom are silicide and gate oxide regions respectively (the silicide region is further indicated by the arrow in Figs. 10.11(b)–(f)). The target ROI for this sharpening process is the gate/gate oxide/implant volume which requires the tip apex to reside near the silicide region and the diameter at the gate oxide to be less than 200 nm.

Tip shaping is accomplished by applying an annular milling pattern (top of Fig. 10.11(b)) with constant outer diameter ($\sim 4\,\mu$m) and a decreasing inner diameter (Figs. 10.11(b)–(d)) with a beam current of 0.28 nA. The milling proceeds from the outer diameter to the inner diameter of the pattern. The first pattern (Fig. 10.11(b)) has an inner

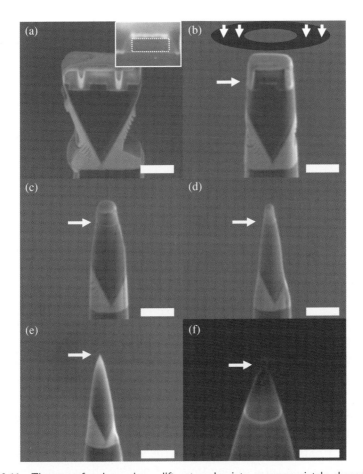

Fig. 10.11   The steps for sharpening a lift-out wedge into an appropriately shaped atom probe specimen. (a) Original wedge before tip shaping with a close up of the transistor region contained within the wedge (inset-dashed box highlights the gate region of the transistor). (b) The first annular milling pattern (top gray halo with downward arrows indicating ion beam direction) shapes the tip into a cylinder with the silicide region of the transistor indicated for clarity (arrow in all succeeding images). (c) Second milling pattern producing a slightly tapered and narrowed end. (d) Third milling pattern producing a more tapered and narrowed end. (e) Final tip shape after 5 kV ion clean-up step. (f) High magnification image of final tip showing the region of the tip for analysis. Note scale bars are 1 μm (a–e) and 200 nm (f).

diameter of $\sim$1.6 μm and produces a long ($>$5 μm) cylindrical shape that proceeds well beyond the Pt-weld region of the mount. The second and third patterns (Fig. 10.11(c) and 10.11(d)) have inner diameters of $\sim$0.6 μm and $\sim$0.3 μm, respectively. The goal of each of these patterns is to push the tip diameter of the inner radius of the pattern to just beyond the bottom of the ROI. The final tip shaping is accomplished with the low-energy (5 keV) FIB clean-up step described below.

As mentioned previously, the implant and damage region created in silicon by a 30 keV gallium ion beam has been shown to extend to ~30 nm into the surface of the sample while 2–5 keV ions limit damage to less than 5 nm. The goal of the low-energy (2–5 keV) clean-up step in this case is to remove the 30 keV damage region, place the apex of the tip at or below the silicide region, and narrow the bottom of the ROI to 200 nm or less in diameter. A $>4\,\mu$m diameter circular milling pattern is centered over the tip and milling proceeds at a reduced beam current (48 pA in the current example). The actual diameter of the pattern and beam current can be adjusted to slow the rate of milling so that the user is able to carefully control the stopping point. Many FIB instruments provide live viewing of milling under such conditions which enables real-time end-point control. Final tip shapes are shown in Figs. 10.11(e) and 10.11(f). In the highly magnified image, Fig. 10.11(f), the silicide region no longer shows bright contrast while the gate oxide region is clearly visible below the center of Fig. 10.11(f). The small cylindrical shape of the silicide region compared to the gate region directly below is a consequence of the differential sputtering rate between the two materials. The pre-low-energy-clean-up tip dimensions must be chosen in anticipation of this behavior. The total milling time to remove the ~500 nm of remaining Pt strip and material above the silicide region is ~60 seconds. Small variations in total milling time can quickly change the dimensions of the final tip shape. For example, an additional five seconds of milling in this case would push the tip apex below the silicide region.

10.4.5. FIB Deprocessing

In the preceding sharpening example, the necessary removal of a large amount of material (~500 nm) to reach the ROI inhibits the user's ability to make a repeatable tip shape because of the influence of differential sputtering due to interaction with grain boundaries and multiple material regions. It is therefore desirable to control the amount of material above the ROI before initiating the tip-shaping procedure. Minimizing the capping material thickness is helpful in this regard, but full control of the ROI-to-surface distance is ideal. Likewise, a problematic region may exist near the ROI, and removing that region by allowing the low energy clean-up step to place the tip apex below that region does not result in an appropriately shaped ROI volume. In this case, complete or partial removal of that region prior to FIB processing and addition of a new sacrificial capping layer can enable proper shaping of the ROI volume.

Both of the above objectives can be accomplished via FIB deprocessing. This process involves removing layers of material parallel to the original sample surface using FIB ion beam milling in a fashion very similar to TEM lamella production techniques.[120] An extracted wedge of material, like that in Fig. 10.10, is first rotated by 90° along its long axis enabling milling parallel to the original wedge surface. This rotation is usually accomplished by transferring the wedge to a manipulator that has an axial-rotation capability.[116] Similar to final APT tip shaping, the SEM is used to endpoint while performing FIB deprocessing of semiconductor devices.

Figure 10.12(a) shows a highly magnified image of a transistor exposed at the edge of an extracted wedge that has been rotated 90° with the full exposed edge of the wedge shown in Fig. 10.12(b). Approximately 500 nm of material must be removed to locate the top surface near the gate oxide. The user can carefully trim any amount of material from the surface. Figure 10.12(c) shows the stopping point one would choose if the top of the silicide region was the goal of the deprocessing while Fig. 10.12(d) shows the stopping point just above the gate oxide.

Once the extra material or problematic material has been removed, the wedge is rotated back to its original orientation and a desirable capping material of appropriate thickness is placed over the new top surface to protect it during the shaping procedure.

10.4.6. Chemical Deprocessing

It is also possible to remove problematic materials with other various techniques including wet chemistry.[121] Chemical deprocessing techniques include wet chemical etching, reactive ion etching, chemical mechanical polishing, and combinations of these techniques commonly

Fig. 10.12 An example of FIB deprocessing. (a) A rotated image of a transistor region within the wedge. (b) The original extracted wedge after it has been transferred to a manipulator that has been rotated 90°. (c) Wedge after it has been deprocessed to a point near the silicide region above the transistor. (d) Wedge after it has been further deprocessed to a point just above the gate oxide region. Note scale bars are 500 nm (a) and 1 μm (b–d).

used in failure analysis. For example, materials that may be difficult to successfully field evaporate, such as SiO_2 and SiN, can be removed with applications of dilute hydrogen fluoride and heated phosphoric acid respectively. Developing recipes for specific device applications can be very resource intensive and may require TEM analysis to verify complete removal of undesired materials prior to further processing.

10.4.7. Cross-Section and Backside Orientations

Performing analysis in directions other than from the top surface into the bulk of the sample can have advantages. For example, analyzing thin

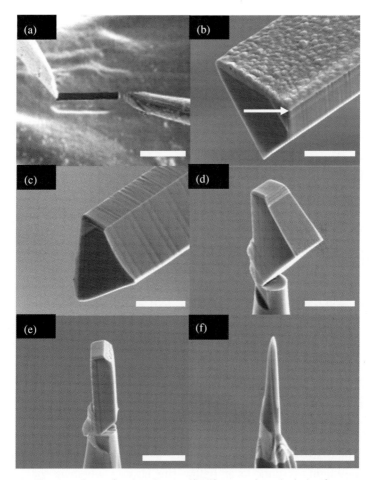

Fig. 10.13   The steps for performing site-specific lift out and manipulation for cross-section APT. (a) The original capped wedge is transferred to a manipulation probe which allows rotation of the wedge. (b) Original wedge still in normal orientation is FIB cut to expose cross-section ROI (see arrow) (c) Wedge is rotated by 90 degrees and has the cross-section ROI surface capped for final preparation. (d) A wedge section is mounted to a post. (e) The mount is trimmed prior to annular milling. (f) The final tip with cross-section ROI captured at the tip. Note scale bars are 25 μm (a) and 3 μm (b–f).

films parallel to the film interface orientation (termed "cross-section" in the current work) serves to both increase the volume of the ROI (the film interface) and improve the analysis yield (number of successful analyses without premature specimen failure) by changing the orientation of the applied stress relative to the interface(s). Figure 10.13 illustrates a cross-section specimen preparation procedure where the ROI consists of layers of material both at and near the original specimen/vacuum interface.

10.4.7.1.   Cross-section preparation

As mentioned previously, a thick sacrificial cap is first added to the surface of the ROI to both protect the ROI and bulk up the surface layer so that the near-surface ROI can be centered in the final tip.[116] In this case, ~500 nm of Ni cap has been added to the top surface before a standard lift-out procedure is initiated to carve out a larger-than-normal wedge (~5 μm wide). Note that it may be beneficial to choose the capping layer material to have similar ion milling and field evaporation properties as to the substrate material. In addition, this capping layer should be a single element or alloy. This additional thickness, with respect to the standard lift-out process described above, is necessary because the capping material may comprise up to half of the final specimen shape, Fig. 10.13.

After attaching the manipulator and extracting the wedge from the bulk, the wedge is transferred to a horizontal manipulator with an axial-rotation degree of freedom (Fig. 10.13(a)). A tip is transferred by first attaching the wedge to the second manipulator with FIB Pt and then cutting the original manipulator from the wedge. While attached to the horizontal manipulator, one side of the wedge is milled flat, creating a new surface 90° to the original surface as shown in Fig. 10.13(b). The arrow highlights a dark layer below the Ni cap which is the location of the ROI in this example. For site-specific cross-section preparation, the location of this new surface is critical. In that case, the new surface needs to be positioned directly above the feature of interest. Because this example is a 2D thin layer, the location of the new surface is somewhat arbitrary.

Figure 10.13(c) shows the wedge after it has been rotated 90° and an additional capping layer (~50 nm Ni) has been added to the new surface. Sufficient capping material on the new surface is especially important in this case because of the propensity for Ga ions to channel along material interfaces. The strategy here is to place enough capping material over the new surface so that the tip can be shaped with 30 kV ions and cleaned-up with 5 kV ions without exposing the material interfaces to Ga ions. Without the residual capping material, even the 5 kV Ga ions may channel many tens of nm into the ROI. This channeling can also result in severe consequences for the tip shape. Should the cap be removed before tip shaping is complete, differential milling of the different layers can cause the tip to form into multiple tips negating its utility for APT analysis.

Once the capping material has been added, the wedge is trans-ferred back to the original micromanipulator and propagated to carrier tips in a manner similar to the standard lift-out method (Fig. 10.10). Figure 10.13(d) shows a portion of the wedge after it has been success-fully propagated to a carrier tip. Before tip shaping can begin, effort is made to make the tip more cylindrical. The portion of the wedge that extended beyond the edge of the microtip is removed with a rectangular milling pattern before sharpening (Fig. 10.13(e)). The final tip shape after annular milling and low-energy clean-up is shown in Fig. 10.13(f).

Because the composition (and thus the ion mill rate) of the specimen does not have cylindrical symmetry, the annular milling process used to shape the tip may provide new challenges. Any sputtering differential between layers will cause asymmetry in the final shape. Often the center of the milling pattern will need to be positioned away from the center of the ROI in anticipation of one side of the specimen milling at a higher rate. The proper amount of shift is usually determined by trial and error experience. Choice of the capping material used to bulk up the original surface to have similar milling properties of the ROI will mitigate this phenomenon.

10.4.7.2. Backside preparation

Analyzing materials from within the bulk towards the surface (termed "backside" orientation in the current work) can serve multiple pur-poses.[87,117] Should weak or otherwise problematic materials or inter-faces exist between the sample surface and the ROI, analysis from the backside may allow for these regions to be avoided. In addition, tip dis-tortions and consequently atomic reconstructions can be affected by the order of evaporation of regions with differing evaporation field require-ments. Performing analysis both from the top-down and backside can provide meaningful information to help separate real observation from artifact.[58,61]

Backside specimen preparation requires application of sufficient material to the top surface of the specimen so that the overall depth of the wedge remains ~2–3 μm after it has been rotated 180° and shaped into a sharp specimen. As in previous examples, adhesion of the added material to the original sample surface is a concern because that interface will reside near the apex of the final specimen. One successful recipe used a palladium seed layer followed by a ~3 μm silver layer to provide the necessary thickness for the backside procedure.[61,117] The palladium provided good adhesion between the top metallic surface and the silver, while the silver provided a high sputtering rate allowing for fast deposition of 3 μm of material. The actual steps of the backside specimen preparation are similar to those for cross-section.

10.4.8.   Hybrid TEM/APT Preparation

TEM and APT have different advantages and limitations, but together they provide complementary information to produce reliable characterization of the three-dimensional position and chemical identity of the atoms in any materials system. Although the FIB/SEM provides some information about the structure of an APT specimen after the specimen fabrication process, higher-resolution characterization of specimens using TEM and STEM are useful to further increase APT reconstruction accuracy. TEM can image specimen radius and shank angle with high precision and give internal structure of interfaces and precipitates. Analytical techniques such as EDS and EELS as well as STEM-HAADF imaging can provide preliminary information about composition of precipitates and interfaces. A number of groups have reported progress in developing general hardware and methods that allow straightforward TEM and APT analysis of the same specimen.[57,58,86,87,117,122–124]

Comparison in three dimensions of the same needle-shaped specimen using electron tomography (ET) and APT has significant advantages. ET provides larger-scale morphological information and APT provides smaller-scale detailed compositional information. Arslan et al.[86] were able to demonstrate good 3D agreement for the same volume of Ag-Al particles with both techniques, without scaling one method to the other, illustrating the high resolution and spatial accuracy of *both* techniques. Such comparisons allow for direct confirmation of artifacts theoretically known to be present in both techniques, which aids in understanding the evaporation process in APT, determining optimal reconstruction parameters for both techniques, and evaluating the quality of the ET and APT reconstructions.

Because neither technique can unambiguously identify the true atomic identity and coordinates of all the atoms in a volume — the ultimate goal of atomic-scale microscopy — a new Atom TOMography (ATOM) concept and project is currently being proposed.[88] In this atomic-scale tomography concept, the APT and STEM approaches are combined into a single instrument called the atomscope. This combination would permit the full characterization of the microstructure of a specimen to be made at the atomic level. Not only could the local crystallography and solute distribution for all elements be evaluated but the three-dimensional data could be used to generate many other types of information, such as the mechanical and electrical properties of the material. The atomscope could be used in three main modes of operation in addition to stand-alone conventional operation as a STEM and a LEAP. The first mode would be to perform electron tomography followed by APT. The second mode would be a time-sliced method to perform sequences of electron imaging, including high resolution electron microscopy and/or surface profiling, then APT. The third mode would be to do simultaneous electron imaging and APT.

10.4.9. Applications

"Atom probe tomography" does not simply refer to an atom probe instrument; it encompasses the instrumentation, data analysis software, operational protocols and finally the rigor of the analyst as well as her, or his, design of experiment. APT can be used to analyze microelectronic structures which vary widely in complexity. The way resulting data are used can also vary widely and, as with all analytical techniques, their usefulness depends upon the confidence the analyst has in the measurement. The question, "Are these two specimens different?" cannot be answered simply by collecting data and looking for differences. The precision of the measurements must be understood first, before the difference question can be adequately addressed.

10.4.10.   Precision and Accuracy

Precision, sometimes referred to as reproducibility or even repeatability, is the degree to which repeated measurements yield the same result.[125] It is important to understand the statistical variation in the data from multiple analyses of a single specimen before one can confidently state whether or not the data from two or more specimens indicate a statistically significant difference. Since APT is destructive, it is never possible to analyze a single specimen more than once. It is therefore necessary to test an atom probe instrument's precision using a sample which is reasonably believed to be uniform over a volume large enough to provide sufficient material to manufacture a population of "identical" atom probe specimens.

Before examining precision of a population of analyses, self consistency of a single analysis should be considered. During an atom probe analysis, the specimen becomes blunter as the analysis proceeds. The applied voltage is gradually increased accordingly in order to maintain the electric field strength at the tip at the appropriate level for pulsed field evaporation. So, a number of analysis variables (tip shape, field of view, voltage, etc.) are all changing during the analysis bringing into question the precision or self consistency of a measurement within a single atom probe specimen. In order to illustrate this consistency, a specimen prepared in cross-section (as described above, and shown in Fig. 10.14) was reconstructed over four voltage ranges within a single run. Figure 10.15 shows the results of the self-consistency experiment. The sample analyzed was from a plain (i.e. un-patterned) Si wafer, implanted with a As and B and reasonably assumed to be uniform in composition over the volume from which the atom probe tip was lifted out.

Within statistical error, the data are the same for the four reconstructed ranges. This is an important result because, without this confidence in self consistency, tip-to-tip comparison would be valid only if

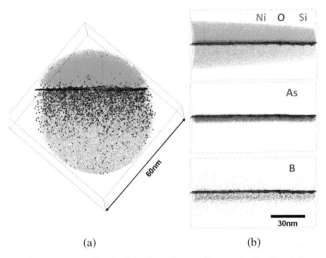

(a)　　　　　　　　　　　　(b)

Fig. 10.14　APT atom maps of a dual-implant As-and-B sample showing (a) an *xy* view and (b) *xz* views of the individual atom species.

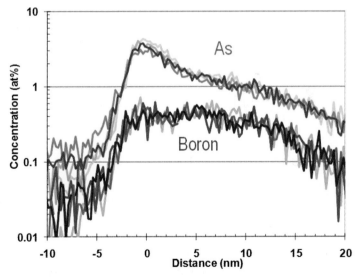

Fig. 10.15　Composition profiles for implanted As and B reconstructed using the same parameters over four different voltage ranges within a single run. No statistically significant variation can be observed in the overlaid data, which shows good self consistency of the measurement within a single analysis.

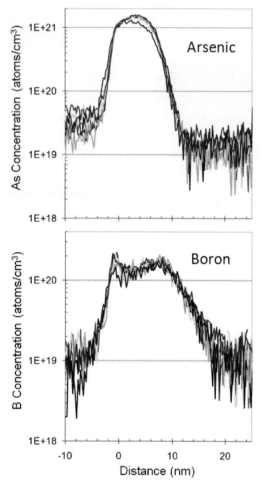

Fig. 10.16   Composition profiles for arsenic and boron showing run-to-run precision for six individual specimens.

the region of interest had the same voltage evolution (was at exactly the same position in depth for every tip with identical tip shapes). This is not the case and tip-to-tip comparisons can now be confidently examined.

Figure 10.16 shows the results from multiple atom probe analyses from a different dual-implant (again As and B) plain Si wafer. A population of six atom probe tips was made from a single FIB lift-out wedge (similar to that shown in Fig. 10.10) and analyzed in cross-section using identical instrument and reconstruction parameters.

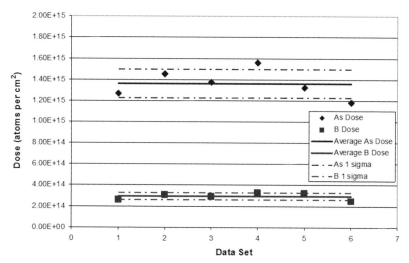

Fig. 10.17   Arsenic and boron dose calculations for the data shown in Fig. 10.16. The one sigma relative standard deviation is better than 10% for these data.

Figure 10.17 shows the results of the B and As dose calculations from the six datasets. The one-sigma relative standard deviation (RSD) was better than 10%. This number is important because now we have determined a criterion to distinguish whether two subsequent measurements of B and/or As are statistically the same or genuinely different. Note that these analyses were done in cross-section mode and the datasets contain larger numbers of ions, with a lower signal-to-noise, than typical top-down analyses. The purpose of this exercise was to demonstrate that atom probe systems are sufficiently stable and analytical protocols are sufficiently robust to enable 10% RSD measurements to be performed in a routine manner.

It has been demonstrated that atom probe instrumentation is capable of statistically precise measurements, but it should be noted that even though a measurement is precise, it does not automatically follow that it is accurate — the quality of being near the true value.[125] Standard reference materials (SRM) can be obtained from the US National Institute of Standards and Technology (NIST) and used to evaluate the accuracy of a measurement. NIST SRM 2133, 2134, and 2137 are the individual samples typically used for calibration of a metrology instrument for phosphorus, arsenic, and boron dopant, respectively. The results of repeated APT analyses (four each, in top-down orientation) of these three samples are shown in Fig. 10.18 for (a) boron, (b) phosphorus, and (c) arsenic, with the thin blue line in each profile showing the NIST measurements provided with the SRMs. Estimates may be obtained for (1) accuracy, by comparing the absolute levels of the APT results to the NIST SIMS values (blue line), (2) sensitivity by measuring the concentration value at which the APT data become nominally flat, and (3) precision, by comparing the individual APT measurements against one another in a single profile.

The best overall results are shown for the arsenic specimens. For the boron, measured values of concentration fall below the NIST SRM value, which is likely due to the fact that some boron atoms not being detected in near-simultaneous multiple defects events. The phosphorus measurements show reduced sensitivity compared to the others, which is likely due to the fact that the phosphorus peak is located only slightly higher in the mass spectrum than the main silicon peak. Note that these results are presented in a "raw data" state with no normalization parameters employed.

In summary, as an introduction to this section on microelectronics applications, the atom probe "technique", which comprises specimen preparation, instrumentation, data analysis software, as well as operational and data analysis protocols, can be demonstrated to be both precise and reasonably accurate for straightforward analyses, such as those described above. Without such a demonstration of the accuracy and precision for straightforward samples, it would not be possible to have confidence in the reliability of data from the more complex structures presented in the remainder of this chapter.

10.5. SEGREGATION IN SILICON

The controlled placement of dopant atoms in Si is critical in the fabrication of semiconductor field-effect devices where each electrically "active" dopant atom contributes an electrical carrier to the lattice. A non-homogeneous distribution of dopant atoms on the nanoscale regime has the potential to create a corresponding fluctuation in the electrical characteristics of that region. In macro-sized devices, these localized

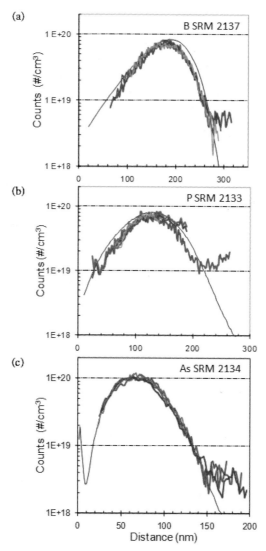

Fig. 10.18   Comparison of four different APT measurement results for implantation of (a) boron (SRM 2137), and (b) phosphorus (SRM 2133), and (c) arsenic (SRM 2134) with the NIST-provided measurement shown as the thin blue line in each profile.

fluctuations average out over the relatively large area of the device and therefore have a negligible impact on overall performance. In nanoscale devices, however, localized fluctuations have an increasingly important impact on the electrical characteristics of individual devices. Imaging of

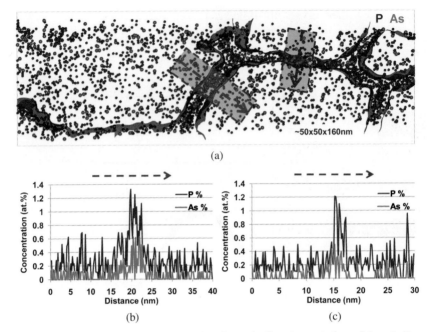

(a)

(b)   (c)

Fig. 10.19   (a) APT atom map showing phosphorus (red) and arsenic (green) in polysilicon
with an isoconcentration surface of (P + As) > 0.6 at.% also shown. (b, c) show compositional analysis of the regions denoted by the dashed boxes in (a). Both analyses show
significantly different levels of As and P at the grain boundaries as compared to the matrix
levels (data binned to 0.2 nm blocks).

individual dopant atoms is essential to the development of advanced
nanoscale devices and is therefore an area of intense interest within the
semiconductor community.

An atom map of polysilicon doped with phosphorus and arsenic
atoms is shown in Fig. 10.19(a). The entire analyzed dataset is $\sim$50$\times$
50 $\times$ 160 nm^3 and the grain size in the silicon is estimated to be
47 $\pm$ 13 nm. A nonhomogeneous distribution of the dopant atoms is
clearly illustrated from the data. Segregation of both dopant species to
the grain boundaries is quantified in Figs. 10.19(b)–(e) for the two grain
boundary regions outlined by dashed boxes in Fig. 10.19(a). The figures show an enrichment at the grain boundaries to between 1–1.4 at.%
P and between 0.2–0.4 at.% As, as compared to the matrix levels of
$\sim$0.2% and <0.1%, for the P and As, respectively. Note that the level
of detected As is below the bulk As solubility limit (at 900°C) in both

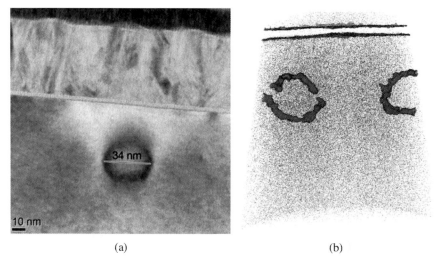

(a)   (b)

Fig. 10.20   (a) Transmission electron micrograph and (b) APT atom map of dislocation loops in As-implanted silicon. The image in (b) shows an isoconcentration surface enclosing the volume of material containing at least 2 at.% As. (Note that the scale is the same in both images.) Figure from Ref. 127.

matrix and at grain boundaries, but the P level appears to exceed the bulk solubility limit (at 900°C) at the grain boundary.

Another example of inhomogeneous dopant distributions in silicon is shown in Fig. 10.20, which shows a Cottrell atmosphere[126] surrounding a dislocation. Figure 10.20(a) shows a TEM image[127] for an As-implant sample after annealing at 600°C for 0.5 hr followed by 1000°C for 30 s. The end-of-range defects observed prior to high-temperature annealing have completely dissolved, and have transformed into dislocation loops in the As-rich region of the sample. APT data from the same region of specimen are shown in Fig. 10.20(b), where two dislocation loops are seen to be decorated by As atoms (As isoconcentration surfaces containing the volume of material having >2 at.% As are shown in the figure).[127] These dislocation loops had been observed previously with TEM but the Cottrell atmospheres had never been seen in silicon prior to the observation with APT.

The impact that these Cottrell atmospheres may have on the concentration of electrically active dopant atoms — and therefore on device performance — remains undetermined but could prove important. Consider the smallest device on a processor manufactured using older 90-nm technology. This device has channel lengths on the order of 50 nm. The Cottrell atmospheres measured above indicate a factor of 10 increase in As concentration over a ~3-nm region. It is therefore unlikely that the localized dopant concentration increase caused by the As Cottrell atmosphere would cause a noticeable electrical impact in a 90-nm device. Now consider the 22-nm technology node, with minimal features on the order of ~10 nm. The presence of a single-dopant Cottrell atmosphere within such a device could alter the electrical performance of the individual device in a way that renders it nonfunctional, or at least renders it dissimilar to an identical device that did not contain a Cottrell atmosphere.

10.6.   SILICIDES

NiSi currently is the best silicide for contact material in advanced integrated circuits but its high-temperature stability is of concern.[128] It has been shown that the addition of Pt in the Ni film increases the nucleation temperature of $NiSi_2$ by approximately 150°C and thus stabilizes the NiSi films.[129] The addition of a small amount of Pt also increases the temperature of agglomeration of NiSi.[130] Ni(Pt)Si thin film alloys are thus currently used as contact layers in complementary metal-oxide-semiconductor devices to improve the integration of these devices in nanoscale transistors.[131] The formation behavior of Ni silicide alloyed with Pt has recently been analyzed by APT,[132–136] two examples of which are contained in this section. APT atom maps from a NiSiPt structure formed by annealing of a 50 nm thick Ni-5 at.% Pt film sputter-deposited onto Si are shown in Fig. 10.21.

The annealing was halted prior to total consumption of the NiPt layer[137] in order to simultaneously form NiSi (low resistivity) and Ni_2Si phases. Figures 10.21(a)–(b) show the location of the Ni, Si and Pt atoms.[135] Pt enrichment is observed at the top of the structure (corresponding to the unreacted Ni layer) and at the interface between silicide and silicon regions. A schematic of the structure is shown in Fig. 10.21(c), which illustrates the regions of the different silicide phases. By taking a thin slice in the vertical direction of Fig. 10.21, the Pt

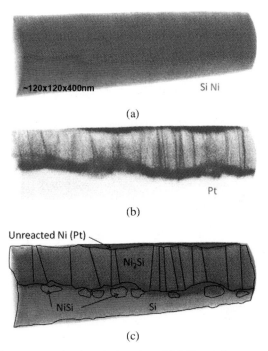

Fig. 10.21   APT atom maps showing the location of (a) Ni and Si atoms and (b) Pt atoms in a NiSiPt thin film formed by annealing a NiPt alloy deposited onto Si. (c) Schematic representation of the various phases that have formed during the reaction. Figure from Ref. 135.

distribution within both Ni$_2$Si (Fig. 10.22(a)) and NiSi (Fig. 10.22(b)) regions may be investigated. Pt segregation to grain boundaries in the Ni$_2$Si region is seen in Fig. 10.22(a), while the NiSi layer shown in Fig. 10.22(b) is seen to be discontinuous with high levels of Pt contained between the NiSi grains, Fig. 10.22(c) (8 at.% Pt isoconcentration). This analysis[135] shows that the reaction of a 50 nm thin film of Ni-5 at.% Pt with Si(100) substrates results in the simultaneous presence of NiSi grains and Ni$_2$Si layer together with the Ni alloyed layer. The NiSi layer is not homogeneous and the grain sizes of the NiSi phase are greater than those of the Ni$_2$Si phase.

   Kim et al.[133] and Adusumilli et al.[136,138] have used APT to map distributions of Pt or Pd in Ni monosilicide. Solid-solutions of Ni$_{0.95}$M$_{0.05}$ (M = Pd or Pt) thin films on Si (100) substrates were subjected to rapid thermal annealing (RTA) to form the monosilicide phase. In both

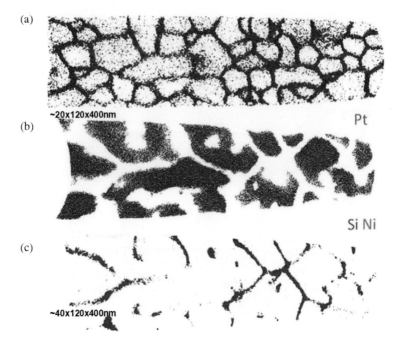

Fig. 10.22   Selected slices from data shown in Fig. 10.21 which illustrate (a) Pt segregation in the Ni_2Si phase, (b) Si and Ni atoms in the discontinuous NiSi phase, and (c) Pt segregation (isoconcentration surface at 8 at.% Pt). Note the registration between the high content Pt regions in (c) and the low concentration of Ni+Si regions in (b). Figure from Ref. 135.

the cases, the transition metal alloying element, Pt or Pd, is found to segregate at the nickel silicide/silicon heterophase interface. The Gibbsian interfacial excesses of Pd and Pt at the silicide/silicon heterophase interface were found to be 3.4 ± 0.2 atoms nm^{-2} and 1.2 ± 0.1 atoms nm^{-2}, respectively. A measured decrease of the interfacial Gibbs free energy due to the segregation at the silicide/silicon interface is most likely responsible for stabilization of the Ni monosilicide phase at elevated temperatures. Furthermore, quantitative evidence for short-circuit diffusion of Pt via grain boundaries (GBs) in the NiSi phase is observed in 3-D direct space, providing valuable insights into the kinetics of this reactive diffusion process.

Figure 10.23(a) displays a 5 nm thin slice of APT data from $Ni_{0.95}Pt_{0.05}/Si(100)$ subjected to RTA at 420°C for five seconds. Each dot represents an individual atom and the different elements are color

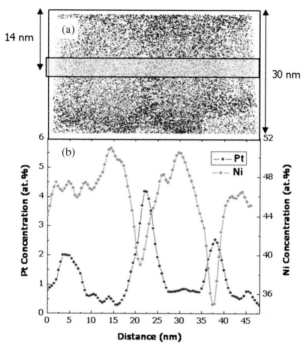

Fig. 10.23 (a) APT atom map in NiPtSi alloy ($30 \times 50 \times 5$ nm^3) and (b) composition profiles of Pt and Ni atoms within the $4 \times 4 \times 48$ nm^3 cylindrical volume element placed horizontally across three grain boundaries of the ($Ni_{0.99}Pt_{0.01}$)Si grains as displayed in (a). Figure from Ref. 136.

coded: Nickel, silicon and platinum atoms are represented by green, blue, and red dots, respectively. Pt is present along the GBs of the nickel monosilicide grains and at the interface between NiSi and the Si(100) substrate. Figure 10.23(b) displays the one-dimensional concentration profiles of Ni and Pt across three $Ni_{0.99}Pt_{0.01}Si$ grain boundaries. At the grain boundaries, an observed increase in the Pt concentration and a concomitant decrease in the Ni concentration points to short-circuit diffusion of Pt along the grain boundaries in the monosilicide. The high spatial resolution coupled with the unique 3-D nature of the measurements yields accurate and precise measurements of both the lattice ($D_L = 4.2 \pm 1.2 \times 10^{-19}$ m^2/sec) and grain boundary ($D_{GB} = 1.5 \pm 0.2 \times 10^{-16}$ m^2/sec) diffusivities of Pt. The ratio of D_{GB} to D_L is $\sim$350. Thus, short-circuit diffusion of Pt via grain boundaries is found to be the predominant mechanism for mass transfer and strongly influences the nickel monosilicide grain size, growth, and morphology. Lattice diffusion of Pt into the bulk from the top metal/silicide interface, short-circuit diffusion via the grain boundaries, and lattice diffusion into the grains from the grain boundaries indicate a physical situation that is analogous to Harrison's Type B diffusion model.[139]

10.7. HIGH-κ DIELECTRICS

The gate dielectric is a critical component of any MOS device. The 32-nm (and beyond) nodes of the International Technology Roadmap for Semiconductors project effective oxide thicknesses that are less than 1 nm thick. This can be accomplished only with a dielectric material whose permittivity exceeds that of silica. Hafnia-based dielectric films, with a dielectric permittivity at least five times greater than that of silica, have recently replaced SiO_2 as the gate dielectric of choice for Si-based CMOS device technology beyond the 45 nm node. As the target effective oxide thickness (EOT) of gate dielectrics approaches 1 nm for SiO_2, the corresponding HfO_2 layer thickness achieves the same capacitance at a thickness of 5 nm. The advantages of the thicker hafnia-based layer include a reduction in gate leakage current and a more robust process. The electrical performance of high-κ materials is critically dependent on the physical properties of the deposited film. Of particular interest are the chemical composition (i.e. density and stoichiometry) and the interface integrity (i.e. interface roughness and chemical interdiffusion within adjacent materials) of the deposited film. APT has been applied to the analysis of Hf-based high-κ dielectrics[140,141] and also to a high-κ layer consisting of HfO_2 with an embedded sub-nm thick ZrO_2 layer.[142,143]

Figure 10.24 shows a scanning electron micrograph of a typical tip before the analysis along with the reconstructed atom map and the cylinder used for extracting composition profiles.[142] The cylinder was aligned in the center of the reconstructed volume to exclude regions which may have been damaged during the sample preparation.

In this work, Mutas et al.[142] investigated the influence of laser energy and specimen temperature and found that the variation of the laser energy and the specimen temperature have different effects on Silicon and the high-κ layer. While higher laser energies lead to surface

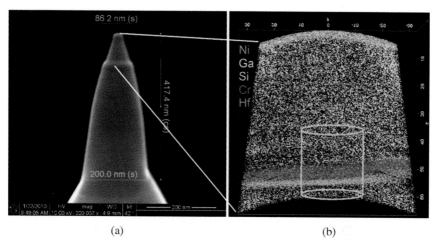

Fig. 10.24 (a) Scanning electron micrograph and (b) APT atom map of Hf-based high-κ dielectric film. Figure from Ref. 142.

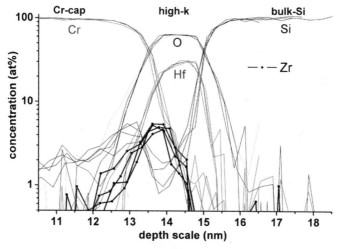

Fig. 10.25 Four APT composition profiles illustrating the repeatability of a Hf-based dielectric film measurement. The structure contains an embedded ZrO_2 layer which is <1 nm in thickness. Figure from Ref. 142.

migration of bulk-silicon, the composition of the high-κ material remains unaffected even at very high laser energies (corresponding to a charge state ratio of $Si^{++}/Si^{+}{\sim}2$) and at specimen base temperatures up to ${\sim}100$ K.[142] Of particular interest in their work is the repeatability of the measurements. Figure 10.25 shows four composition profiles from APT analyses performed at constant specimen temperature (54 K) and laser energy (0.2 nJ).

10.8. COMPOUND SEMICONDUCTORS: QUANTUM WELLS AND QUANTUM RINGS

Gallium nitride (GaN) has achieved great success as the basis for a wide range of optoelectronic devices. Updoped GaN emits in the ultra-violet (UV), and alloying it with indium nitride (InN) or aluminum nitride (AlN) produces emission with wavelengths ranging from the green to the deep UV. GaN-based multiple-quantum-well (MQW) structures are used as the active regions of commercial light-emitting diodes and laser diodes. Recently APT has been used to investigate the chemical and morphological microstructure of GaN- and GaAs-based structures.[144–150] Figure 10.26 (center image) shows ten repeats of

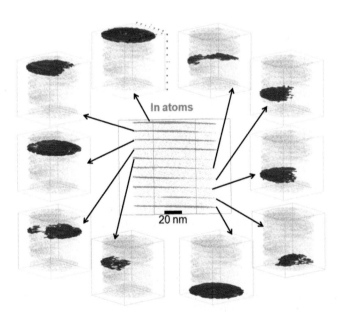

Fig. 10.26   Ten repeats of a UV-emitting $Al_{20}Ga_{80}N/In_xGa_{1-x}N$ multilayer (center image showing In atoms only). Surrounding the center image are isoconcentration surfaces at 0.2 at.% In for each of the individual quantum wells. Both continuous and discontinuous layers are observed.

a UV-emitting $Al_{0.2}Ga_{0.8}N/In_xGa_{1-x}N$ multilayer where the nominal In content, x, is 5 at.%. Photoluminescence measurements[151] determined that the internal quantum efficiency in these structures was 67% at an emission wavelength of 380 nm. The insets surrounding the center image in Fig. 10.26 show isoconcentration surfaces at 0.2 at.% In for each of the individual quantum wells.

Although some of the layers are continuous in the field of view (most notably layers one, three, and ten), several have large gaps and discontinuities on a 20–100 nm length scale. Even in the regions where the layers were present, there are significant variations in the indium content, with the indium content decreasing toward the gaps.[148] This matches well with the uniformity of the layers observed in STEM-HAADF images if care is taken with the imaging conditions.[146] It has been proposed[148] that these 20–100 nm discontinuities in high efficiency UV $In_xGa_{1-x}N$ QW layers form network structures such that threading dislocations pass through the gaps in the $In_xGa_{1-x}N$ layers. This may result in carriers and threading dislocations being separated from one another, explaining the high efficiency seen in the current samples.

Quantum confinement of electrons and holes in separated regions of a nanostructure (type II nanostructures) holds promise for photovoltaic solar cells, opto-electronic devices operating in the infra-red range[152] and memory devices,[153] because of the long hole lifetimes inside the nanostructure.[154] A recent APT study has shown the presence of self-assembled ring-shaped nanostructures in a type II gallium-antimony/gallium-arsenide (GaSb/GaAs) system.[155] Because nano-rings also possess atom-like properties, they are a promising venue for new device applications in optics, optoelectronics, quantum cryptography and quantum computing.[156]

A GaAs/GaSb multilayer structure was grown by MBE on a GaAs (001) substrate and capped with a 300 nm thick GaAs layer.[157] Previous TEM analysis had revealed small nanostructures that formed with heights of ~ 1 nm.[158] These nanostructures were a result of significant Sb segregation that occurred after GaAs capping of the GaSb monolayers.[158] APT analysis[155] reveals an Sb segregation profile where each Sb-rich layer starts with ~ 7 at.% Sb and follows an exponential decay in Sb-concentration corresponding to the depletion of Sb as it is capped

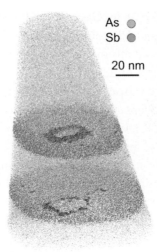

As ●
Sb ●

20 nm

Fig. 10.27   Sb (orange) and As (blue) atom distributions revealing different layers in this GaSb/GaAs system. Red ring-like structures are shown by the red iso-concentration surfaces $(GaAs_{1/3}Sb_{2/3})$. Figure from Ref. 155.

with GaAs. The presence of Sb extends over 10–15 nm above each layer. Figure 10.27 shows the GaSb layers as well as the 20 nm wide ring-like Sb-rich nanostructures. The Sb composition within the rings ranges from 30–36 at.%. The morphology of these nanostructures evolves to a ring shape during capping.

10.9. NANOWIRES

Nanowires have gained enormous attention worldwide due to their unique structural and physical properties and their potential as building blocks for future microelectronic devices[159]; however, their small diameters, extreme aspect ratios, and the variety of as-grown geometries pose challenges for compositional characterization by traditional techniques such as TEM and SIMS.[160,161] Specimen preparation for APT analysis presents its own challenges. In some cases, the individual nanowires can be removed safely (without inflicting ion damage to the analysis volume) and transferred to an appropriate carrier geometry for analysis (see Figs. 10.28(a)–(b) for example),[161,162] but in many cases the density and aspect ratio can make this especially difficult. The most widely

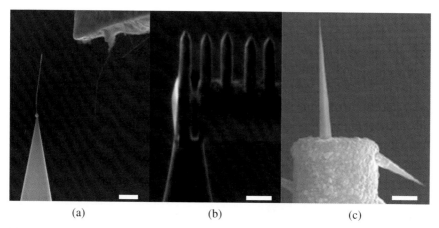

(a)	(b)	(c)

Fig. 10.28   Some methods available for preparation of nanowires for APT analysis. (a) Transfer of free nanowires to a sharp microtip carrier is shown. In this case free nanowires stick to the micromanipulator via static forces and are positioned near to the apex of the sharp microtip and attached with platinum. (b) Transfer of a protected lift-out wedge of nanowires is shown. This process allows for specimen preparation using standard lift-out methods. After the wedge has been propagated, the protective coating and any redeposited material are removed with a combination of FIB deprocessing and wet chemical etching.[161] (c) Nanowires grown in place over a silicon micropost. Figure from Ref. 33.

published and successful approach for studying nanowires with APT has focused on growing the nanowires on carrier tips in a ready-to-analyze geometry (see Fig. 10.28(c) for example).[163–169] This method requires that microposts are first manufactured from a planar growth substrate using available methods (cutting features into a surface with a wafer dicing saw, or etching features using photo lithography and reactive ion etching techniques). Then, catalyst particles can be distributed onto the micropost surfaces followed by nanowire growth.

Dopant distribution and heterojunction analysis are two examples of nanowire compositional characterization uniquely enabled by APT.[33] Figure 10.29 shows the reconstruction of a phosphorus-doped germanium nanowire grown with a gold nanoparticle.[167] The gold catalyst is visible (Fig. 10.29(a)) as are the atomic planes normal to the $\langle 111 \rangle$ growth direction (Fig. 10.29(b)). The doping rate, which is not generally known, is determined by the rate at which the dopant atoms move from the gas phase as precursors to the solid phase as substitutional impurities. Quantitative analysis of the mass spectrum showed that the

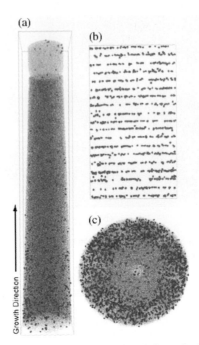

Fig. 10.29 Three-dimensional reconstruction of doped Ge nanowire showing Au, Ge, and P atoms as gold, blue, and gray spheres respectively. (a) Side view of Au-catalyst-tipped nanowire. Arrow indicates growth direction. (b) Side view of a center portion of the same nanowire showing {111} planes perpendicular to the growth axis. (c) End view showing a radially non-uniform doping profile due to surface growth. Figure from Refs. 167 and 33.

dopant concentration in the vapor-liquid-solid-grown nanowire was much less than that of the gas phase. Furthermore, uncatalyzed reactions on the nanowire surface introduced dopants at a different rate, leading to the radial variation in doping level seen in the end view (Fig. 10.29(c)).

10.10. ELECTRICALLY INSULATING AND CERAMIC MATERIALS

The use of a pulsed laser in APT analysis has opened the way for the analysis of bulk insulators.[7] As many semiconductor or microelectronic device applications contain regions of insulating or ceramic materials, two examples will be shown in this section on the analysis of such materials.

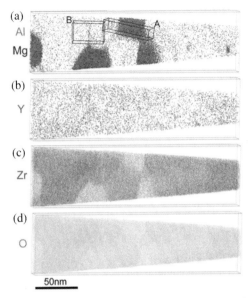

Fig. 10.30   Atom maps of yttria-stabilized tetragonal zirconia with spinel dispersoids. This material is estimated to have an electrical resistivity of $\sim 10^9 \Omega$ cm. Figure from Ref. 35.

Hono *et al.*[35,170] have analyzed bulk insulators from different materials systems (including Ce and Zr oxides). Atom maps of Al + Mg, Y, Zr and O in an yttria-stabilized tetragonal zirconia with spinel dispersoids are shown in Fig. 10.30 ($\sim 15 \times 65 \times 240 \, nm^3$). The presence of nanocrystalline Mg- and Al-rich grains ($MgAl_2O_4$), Fig. 10.30(a) are consistent with the microstructural features observed by the scanning and transmission electron microscopy and Al and Y appear to be segregated along the ZrO_2/ZrO_2 grain boundaries.[35] This material was estimated to have an electrical resistivity of $\sim 10^9 \Omega$ cm. Thus these results, together with the other APT analyses of insulating materials referenced above, illustrate the applicability of APT to entire microelectronic devices, or subsections of such devices.

One such application is in the optoelectronics area ($1.54 \, \mu m$ Si-based planar optical amplifiers), where the interface at the surface of small Si clusters (only a few nanometers in diameter) contained within SiO_2 is expected to dominate the optical properties. Also of interest is the composition, size distribution and number density of the clusters. APT has been used to investigate this system,[172]

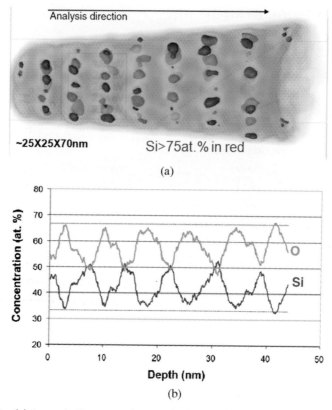

Fig. 10.31   (a) Layered silicon nanoclusters obtained in SiO$_2$. (b) A composition profile along the analysis direction denoted in (a) showing O and Si levels for the individual layers. Figure from Ref. 171.

and the results are shown in Fig. 10.31. In Fig. 10.31(a), color coding is used to denote the percentage of silicon (>75 at.%) and oxygen (>33 at.%) as red and green, respectively. Figure 10.31(b) is the composition profile corresponding to the analysis direction given in Fig. 10.31(a).

This composition as measured by Talbot *et al.*[171] suggests that only a fraction of 50% of the initial silicon excess has precipitated in Si-nanoclusters. They concluded that the annealing time and/or temperature are not long/high enough to ensure a complete phase separation of the system, which agrees with the suggestions of a slow phase separation process in this system.[172,173]

10.11. DEVICE STRUCTURES

One significant goal in the application of APT to microelectronic structures has been the analysis of devices, especially the entire device structure, not just a small portion of a device. The confluence of the downscaling of devices (and the reduction of source/drain junction depths to avoid short channel effects) and the increased field of view now available in modern atom probes makes the analysis of entire device structures a reality. Several researchers have used APT in the analysis of portions of device patterned structures[174–176] and more recently three have succeeded in analyzing entire devices.[33,177,178] Two examples of such device analysis, from a standard transistor and from a fin-style transistor are presented here.

Inoue et al.[177] have investigated 3D dopant distributions in an n-MOSFET (metal oxide semiconductor field-effect transistor) structure, including gate, gate oxide, channel, source/drain extension, and halo, Fig. 10.32 (inset shows TEM image of the analyzed region of

Fig. 10.32   APT atom map of an n-doped MOSFET including gate, gate oxide, channel, source/drain extension, and halo implant. Figure from Ref. 177.

the transistor). Arsenic atoms are visible in the source/drain extension, and are also on top of the gate electrode. B atoms can be seen in the Si substrate (channel and halo extension regions), while P atoms were segregated on the grain boundaries of the poly-Si gate and at the interface between the gate and gate oxide. Phosphorus concentrations of 5×10^{20} cm^{-3} and 2×10^{21} cm^{-3} were measured at the poly-Si grain boundaries and at the gate/gate oxide interface, respectively. The boron concentration in the channel region was measured to be 1×10^{18} cm^{-3} while the As concentration in the source/drain extension was 2×10^{21} cm^{-3} with As atoms being detected to a depth of $\sim$10 nm from the implanted Si surface.

In fin-transistors (FINFETs), device performance is reduced if an inhomogeneous distribution of dopants exists in the source/drain extensions. Recently, Kambham et al.[178] have analyzed the conformality of boron (BF$_2$) implantation in FINFET structures as a function of implantation angle (10° and 45°). Figure 10.33(a) shows the APT atom map of a 40-nm-wide FINFET implanted with B at 10° (angle measured between wafer surface and incoming dopant), where pile-up at the sidewall surface is qualitatively observed. Figure 10.33(b) shows the comparison (both sidewall and top doping profiles) between 45° and 10° implants.[178] For the 45° implantation the top (vertical) peak concentration is twice

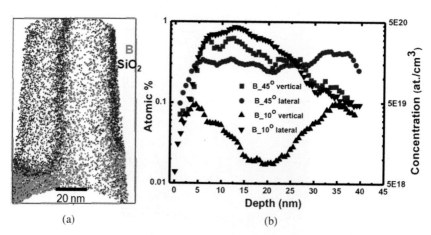

(a)   (b)

Fig. 10.33   (a) APT atom map of boron and silicon dioxide in a FINFET structure and (b) variation in vertical and lateral implantation depth with implant angle (measured between wafer surface and incoming dopant). Figure from Ref. 178.

that of the profile peak concentration. In the $10°$ implantation, the dopants are shallower and significantly reduced in concentration. The result is a high non-conformality of the dopants with the sidewall dose being only 10% of the top dose. These observations were compared favorably with modeling results from Vandervorst *et al.*[179]

10.12. CONCLUDING REMARKS

The transition from micro-electronics to nano-electronics is upon us. Planar transistor technology will be phased out and replaced by three-dimensional device technology over the next few years. This transition will be accompanied by new characterization challenges such as monitoring the dopant distribution versus pattern density in closely spaced structures, measuring top versus sidewall dopant distribution in fin structures, analysis of double gates, complex high-κ gate stack characterization, and as well as additional unforeseen challenges. Designated SIMS structures for proxy process control analyses will be phased out, because they will not be useful for providing information on localized dopant concentration variations. TEM will still be used extensively, but will not be able to provide the "mass spectrometry" and sensitivity required, for dopants and light element distributions. The atom probe can be used to obtain both spatial and compositional information from a nanostructure via a single direct measurement without having to run calibration standards, which makes the technique uniquely suited to meet the demanding new challenges of this dynamic technology sector.

Like all analytical techniques, APT has its deficiencies. Lack of accuracy in reconstruction of polyphase structures is at the top of the list. Many microelectronic structures are polyphase and for this reason, APT has been seen by some as lacking for this class of applications. However, not all applications are heterogeneous polyphase structures, and as shown in this chapter, there are a large number of applications where APT delivers valuable information not available from any other source. This is due in part to the many advantages APT offers for microelectronics applications; very high analytical spatial resolution, very high analytical sensitivity for all elements, and three-dimensional compositional imaging. Ultimately, we believe that the reconstruction accuracy challenges will be solved as well. The future is bright for APT. It is a burgeoning frontier that offers challenges and big rewards to those who make the commitment.

Acknowledgments

The authors would like to acknowledge all of the individuals who contributed to this manuscript, including: P. Ronsheim and P. Flaitz (IBM), E. Marquis (University of Michigan), D. Seidman, P. Adusumilli and L. Lauhon (Northwestern University), A. K. Kambham and W. Vandervorst (IMEC), K. Inoue and Y. Nagai (University of Tohoku), K. Hono (National Institute for Materials Science Japan), C. Humphreys (University of Cambridge), S. Mutas and C. Klein (Globalfoundries), D. Mangelinck and A. Portavoce (Institut Matériaux Microélectronique Nanosciences de Provence), and S. Jin (University of Wisconsin). We would also like to thank the many colleagues at Cameca Instruments Inc., Madison, WI, who assisted in assembling the materials presented in this chapter, including: D. Olson, R. Ulfig, D. Reinhard, S. Gerstl, P. Clifton, J. Olson, D. Lenz, J. Bunton, J. Shepard, T. Payne, E. Strennen, E. Oltman, T. Gribb, D. Rauls, T. Walker, L. Johnson, and J. Watson.

REFERENCES

1. G. E. Moore, Cramming more components onto integrated circuits, *Electronics Magazine*, pp. 114–117 (1965).
2. M. K. Miller, K. F. Russell, K. Thompson, R. Alvis, D. J. Larson, *Microsc. Microanal.* **13**, 428 (2007).
3. J. A. Liddle, A. Norman, A. Cerezo, C. R. M. Grovenor, *Appl. Phys. Lett.* **54**, 1555 (1989).
4. B. Gault, F. Vurpillot, A. Vella, M. Gilbert, A. Menand, D. Blavette, B. Deconihout, *Rev. Sci. Instrum.* **77**, 043705 (2006).
5. G. L. Kellogg, *J. Appl. Phys.* **52**, 5320 (1981).
6. A. Cerezo, C. R. M. Grovenor, G. D. W. Smith, *J. Microsc.* **141**, 155 (1986).
7. D. J. Larson, R. A. Alvis, D. F. Lawrence, T. J. Prosa, R. M. Ulfig, D. A. Reinhard, P. H. Clifton, S. S. A. Gerstl, J. H. Bunton, D. R. Lenz, T. F. Kelly, K. Stiller, *Microsc. Microanal.* **14**(S2), 1254 (2008).
8. C. Oberdorfer, P. Stender, C. Reinke, G. Schmitz, *Microsc. Microanal.* **13**, 342 (2007).

9. A. Cerezo, A. K. Petford-Long, D. J. Larson, S. Pinitsoontorn, E. W. Singleton, *J. Mater. Sci.* **41**, 7843 (2006).

10. A. N. Chiaramonti, D. K. Schreiber, W. F. Egelhoff, D. N. Seidman, A. K. Petford-Long, *Appl. Phys. Lett.* **93**, 103113 (2008).

11. T. F. Kelly, T. T. Gribb, J. D. Olson, R. L. Martens, J. D. Shepard, S. A. Wiener, T. C. Kunicki, R. M. Ulfig, D. Lenz, E. M. Strennen, E. Oltman, J. H. Bunton, D. R. Strait, *Microsc. Microanal.* **10**, 373 (2004).

12. B. Deconihout, F. Vurpillot, B. Gault, G. Da Costa, M. Bouet, A. Bostel, D. Blavette, A. Hideur, G. Martel, M. Brunel, *Surf. Inter. Anal.* **39**, 278 (2007).

13. E. W. Müller, J. A. Panitz, S. B. McLane, *Rev. Sci. Instrum.* **39**, 83 (1968).

14. E. W. Müller, *Z. Phys.* **131**, 136 (1951).

15. E. W. Müller, *Zeitung Technische Physik* **17**(412), (1936).

16. E. W. Müller, T. T. Tsong, *Field Ion Microscopy Principles and Applications*, Elsevier, New York (1969).

17. M. K. Miller, A. Cerezo, M. G. Hetherington, G. D. W. Smith, *Atom Probe Field Ion Microscopy*, Oxford University Press, Oxford (1996).

18. K. M. Bowkett, D. A. Smith, *Field-Ion Microscopy*, North-Holland, Amsterdam (1970).

19. T. T. Tsong, *Atom-Probe Field Ion Microscopy: Field Ion Emission and Surfaces and Interfaces at Atomic Resolution*, Cambridge University Press, Cambridge, Great Britain (1990).

20. M. K. Miller, *Atom Probe Tomography: Analysis at the Atomic Level*, Kluwer Academic/Plenum Publishers, New York (2000).

21. M. K. Miller, G. D. W. Smith, *Atom Probe Microanalysis: Principles and Applications to Materials Problems*, Materials Research Society, Pittsburgh (1989).

22. T. F. Kelly, M. K. Miller, Invited Review Article: Atom Probe tomography, *Rev. Sci. Instrum.* **78**, 031101 (2007).

23. http://www.cameca.com/instruments-for-research/atom-probe.aspx

24. G. L. Kellogg, T. T. Tsong, *J. Appl. Phys.* **51**, 1184 (1980).

25. O. Nishikawa, T. Akimoto, T. Tsuchiya, T. Yoshimura, Y. Ishikawa, *Appl. Surf. Sci.* **76–77**, 359 (1994).

26. T. Kelly, D. J. Larson, *Mat. Char.* **44**, 59 (2000).

27. K. Thompson, D. J. Larson, R. Ulfig, *Microsc. Microanal.* **11**(S2), 882 (2005).

28. J. H. Bunton, J. D. Olson, D. R. Lenz, T. F. Kelly, *Microsc. Microanal.* **13**, 418 (2007).

29. F. Vurpillot, J. Houard, A. Vella, B. Deconihout, *J. Phys. D: Appl. Phys.* **42**, 125502 (2009).

30. A. Vella, F. Vurpillot, B. Gault, A. Menand, B. Deconihout, *Phys. Rev. B* **73**, 165416 (2006).

31. C. R. M. Grovenor, A. Cerezo, *J. Appl. Phys.* **65**, 5089 (1989).

32. R. A. D. Mackenzie, J. A. Liddle, C. R. M. Grovenor, *J. Appl. Phys.* **69**, 250 (1991).

33. L. J. Lauhon, P. Adusumilli, P. Ronsheim, P. L. Flaitz, D. Lawrence, *MRS Bulletin* **34**, 738 (2009).

472 D. J. Larson et al.

34. D. J. Larson, A. Cerezo, J. Juraszek, K. Hono, G. Schmitz, *MRS Bulletin* **34**, 732 (2009).
35. Y. M. Chen, T. Ohkubo, M. Kodzuka, K. Morita, K. Hono, *Scripta Materialia* **6** 693 (2009).
36. T. Ohkubo, Y. M. Chen, K. Nogiwa, A. Nishimura, K. Hono, *Microsc. Microanal.* **15**(S2), 310 (2009).
37. E. A. Marquis, N. A. Yahya, D. J. Larson, M. K. Miller, R. I. Todd, *Materials Today* **13**(10), 42 (2010).
38. T. F. Kelly, O. Nishikawa, J. A. Panitz, T. J. Prosa, *MRS Bulletin* **34**, 744 (2009).
39. T. J. Prosa, M. Greene, T. F. Kelly, J. Fu, K. Narayan, S. Subramaniam, *Microsc. Microanal.* **16**(S2), 482 (2010).
40. Imago Scientific Instruments Corporation, Madison, WI, USA (now part of Cameca SA) .
41. Cameca, S.A., Paris, France.
42. J. H. Bunton, J. D. Olson, D. R. Lenz, D. J. Larson, T. F. Kelly, *Microsc. Microanal.* **16**(S2), 10 (2010).
43. B. A. Mamyrin, V. I. Karataev, D. V. Shmikk, V. A. Zagulin, *Sov. Phys. JETP* **37**, 45 (1973).
44. A. R. Waugh, C. H. Richardson, R. Jenkins, *Surf. Sci.* **266**, 501 (1992).
45. M. R. Scheinfein, D. N. Seidman, *Rev. Sci. Instrum.* **64**, 3126 (1993).
46. A. Cerezo, T. J. Godfrey, S. Sijbrandij, G. D. W. Smith, P. J. Warren, *Rev. Sci. Instrum.* **1**, 49 (1998).
47. S. Sijbrandij, A. Cerezo, T. J. Godfrey, G. D. W. Smith, *Appl. Surf. Sci.* **94/95**, 428 (1996).
48. P. Panayi, *Curved Reflectron* UK Patent No 2427961 (2007).
49. D. Blavette, J. M. Sarrau, A. Bostel, J. Gallot, *Revue Phys. Appl.* **17**, 435 (1982).
50. P. Bas, A. Bostel, B. Deconihout, D. Blavette, *Appl. Surf. Sci.* **87/88**, 298 (1995).
51. B. Gault, F. de Geuser, L. T. Stephenson, M. P. Moody, B. C. Muddle, S. P. Ringer, *Microsc. Microanal.* **14**, 296 (2008).
52. B. Gault, M. P. Moody, F. De Geuser, G. Tsafnat, A. La Fontaine, L. T. Stephenson, D. Haley, S. P. Ringer, *J. Appl. Phys.* **105**, 034913 (2009).
53. B. P. Geiser, D. J. Larson, E. Oltman, S. S. A. Gerstl, D. A. Reinhard, T. F. Kelly, T. J. Prosa, *Microsc. Microanal.* **15**(S2), 292 (2009).
54. S. D. Walck, T. Buyuklimanli, J. J. Hren, *Journal de Physique* **47-C2**, 451 (1986).
55. B. Gault, M. P. Moody, E. A. Marquis, F. De Geuser, B. P. Geiser, D. J. Larson, T. F. Kelly, S. P. Ringer, G. D. W. Smith, *Microsc. Microanal.* **15**(S2), 10 (2009).
56. B. Gault, W. Yang, R. K. Zheng, F. Braet, S. P. Ringer, *Microsc. Microanal.* **15**(S2), 272 (2009).
57. T. C. Petersen, S. P. Ringer, *J. Appl. Phys.* **105**, 103518 (2009).
58. E. A. Marquis, B. P. Geiser, T. J. Prosa, D. J. Larson, *J. Microsc.* **241**, 225 (2010).

59. D. J. Larson, B. P. Geiser, T. J. Prosa, S. S. A. Gerstl, D. A. Reinhard, T. F. Kelly, *J. Microsc.* **243**, 15 (2011).
60. B. Gault, D. Haley, F. De Geuser, M. P. Moody, E. A. Marquis, D. J. Larson, B. P. Geiser, *Ultramicrosc.* **111**, 448 (2011).
61. D. J. Larson, T. J. Prosa, B. P. Geiser, W. L. Egelhoff Jr., *Ultramicrosc.* **111**, 506 (2011).
62. F. De Geuser, W. Lefebvre, F. Danoix, F. Vurpillot, D. Blavette, B. Forbord, *Surf. Inter. Anal.* **39**, 268 (2007).
63. F. Vurpillot, G. Da Costa, A. Menand, D. Blavette, *J. Microsc.* **203**, 295 (2001).
64. B. P. Geiser, T. F. Kelly, D. J. Larson, J. Schneir, J. P. Roberts, *Microsc. Microanal.* **13**, 437 (2007).
65. T. F. Kelly, E. Voelkl, B. P. Geiser, *Microsc. Microanal.* **15**, 12 (2009).
66. B. Gault, M. P. Moody, F. De Geuser, D. Haley, L. T. Stephenson, S. P. Ringer, *Appl. Phys. Lett.* **95**, 034103 (2009).
67. B. Gault, M. P. Moody, F. De Geuser, A. La Fontaine, L. T. Stephenson, D. Haley, S. P. Ringer, *Microsc. Microanal.* **16**, 99 (2010).
68. B. Gault, M. Muller, A. La Fontaine, M. P. Moody, A. Shariq, A. Cerezo, S. P. Ringer, G. D. W. Smith, *J. Appl. Phys.* **108**, 044904 (2010).
69. R. Gomer, *Field Emission and Field Ionization*, Harvard University Press, Cambridge, MA (1961).
70. M. Wada, *Surf. Sci.* **145**, 451 (1984).
71. T. T. Tsong, *Surf. Sci.* **70**, 211 (1978).
72. H. B. Michaelson, *J. Appl. Phys.* **48**, 4729 (1977).
73. D. J. Larson, K. F. Russel, M. K. Miller, *Microsc. Microanal.* **5**(S2), 930 (1999).
74. G. S. Gipson, D. W. Yannitell, H. C. Eaton, *J. Phys. D: Appl. Phys.* **12**, 987 (1979).
75. G. S. Gipson, H. C. Eaton, *J. Appl. Phys.* **51**, 5537 (1980).
76. G. S. Gipson, *J. Appl. Phys.* **51**, 3884 (1980).
77. R. Smith, J. M. Walls, *J. Phys. D: Appl. Phys.* **11**, 409 (1978).
78. T. D. Wilkes, G. D. W. Smith, D. A. Smith, *Metallography* **7**, 403 (1974).
79. A. R. Waugh, E. D. Boyes, M. J. Southon, *Surf. Sci.* **61**, 109 (1976).
80. G. Sha, A. Cerezo, *Ultramicrosc.* **105**, 151 (2005).
81. S. S. A. Gerstl, B. P. Geiser, T. F. Kelly, D. J. Larson, *Microsc. Microanal.* **15**, 248 (2009).
82. A. Shariq, S. Mutas, K. Wedderhoff, C. Klein, H. Hortenbach, S. Teichert, P. Kucher, S. S. A. Gerstl, *Ultramicrosc.* **109**, 472 (2009).
83. M. K. Miller, M. G. Hetherington, *Surf. Sci.* **246**, 442 (1991).
84. E. A. Marquis, F. Vurpillot, *Microsc. Microanal.* **14**, 561 (2008).
85. O. Dimond, *Metallurgy and Science of Materials, Part II Thesis*, University of Oxford (1999).
86. I. Arslan, E. A. Marquis, M. Homer, M. A. Hekmaty, N. C. Bartelt, *Ultramicrosc.* **108**, 1579 (2008).
87. B. P. Gorman, D. Diercks, N. Salmon, E. Stach, G. Amador, C. Hartfield, *Microscopy Today* **16**, 42 (2008).
88. M. K. Miller, T. F. Kelly, *Microsc. Microanal.* **16**(S2), 1856 (2010).

89. B. P. Geiser, D. J. Larson, S. S. A. Gerstl, D. A. Reinhard, T. F. Kelly, T. J. Prosa, J. D. Olson, *Microsc. Microanal.* **15**(S2), 302 (2009).
90. A. J. W. Moore, *Philosophical Magazine A* **43**, 803 (1981).
91. M. K. Miller, J. M. Hyde, *11th European Congress on Electron Microscopy*, Dublin, Ireland (1996).
92. F. Vurpillot, A. Bostel, D. Blavette, *J. Microscopy* **196**, 332 (1999).
93. F. Vurpillot, A. Bostel, A. Menand, D. Blavette, *Eur. Phys. J.-Appl. Phys.* **6**, 217 (1999).
94. F. Vurpillot, A. Bostel, D. Blavette, *Appl. Phys. Lett.* **76**, 3127 (2000).
95. D. Blavette, F. Vurpillot, C. Pareige, A. Menand, *Ultramicrosc.* **89**, 145 (2001).
96. F. Vurpillot, A. Cerezo, D. Blavette, D. J. Larson, *Microsc. Microanal.* **10**, 384 (2004).
97. F. Vurpillot, D. J. Larson, A. Cerezo, *Surf. Inter. Anal.* **36**, 552 (2004).
98. F. Vurpillot, L. Renaud, D. Blavette, *Ultramicrosc.* **95**, 223 (2003).
99. G. L. Kellogg, *Surf. Sci.* **266**, 18 (1992).
100. D. J. Larson, D. T. Foord, A. K. Petford-Long, T. C. Anthony, I. M. Rozdilsky, A. Cerezo, G. D. W. Smith, *Ultramicrosc.* **75**, 147 (1998).
101. D. J. Larson, A. K. Petford-Long, A. Cerezo, G. D. W. Smith, D. T. Foord, T. C. Anthony, *Appl. Phys. Lett.* **73**, 1125 (1998).
102. D. J. Larson, D. T. Foord, A. K. Petford-Long, A. Cerezo, G. D. W. Smith, *Nanotechnology* **10**, 45 (1999).
103. D. J. Larson, A. K. Petford-Long, A. Cerezo, G. D. W. Smith, *Acta Materialia* **47**, 4019 (1999).
104. D. J. Larson, A. K. Petford-Long, Y. Q. Ma, A. Cerezo, *Acta Materialia* **52**, 2847 (2004).
105. D. J. Larson, D. T. Foord, A. K. Petford-Long, H. Liew, M. G. Blamire, A. Cerezo, G. D. W. Smith, *Ultramicrosc.* **79**, 287 (1999).
106. D. J. Larson, B. D. Wissman, R. Martens, R. J. Viellieux, T. F. Kelly, T. T. Gribb, H. F. Erskine, N. Tabat, *Microsc. Microanal.* **7**, 24 (2001).
107. M. K. Miller, *Microsc. Microanal.* **11**(S2), 808 (2005).
108. M. K. Miller, K. F. Russell, G. B. Thompson, *Ultramicrosc.* **102**, 287 (2005).
109. K. Thompson, D. J. Lawrence, D. J. Larson, J. D. Olson, T. F. Kelly, B. Gorman, *Ultramicrosc.* **107**, 131 (2007).
110. M. K. Miller, K. F. Russell, *Ultramicrosc.* **107**, 761 (2007).
111. T. F. Kelly, R. L. Martens, S. L. Goodman, *Methods of sampling specimens for microanalysis*, in U.S.P.a.T. Office (Ed.) U.S. Patent, Imago Scientific Instruments, United States (2003).
112. J. M. Cairney, D. W. Saxey, D. McGrouther, S. P. Ringer, *Physica B* 1 **394**, 267 (2007).
113. J. Takahashi, K. Kawakami, Y. Yamaguchi, M. Sugiyama, *Ultramicrosc.* **107**, 744 (2007).
114. F. Perez-Willard, D. Wolde-Giorgis, T. Al Kassab, G. A. Lopez, E. J. Mittemeijer, R. Kirchheim, D. Gerthsen, *Micron.* **39**, 45 (2008).
115. D. W. Saxey, J. M. Cairney, D. McGrouther, T. Honma, S. P. Ringer, *Ultramicrosc.* **107**, 756 (2007).
116. D. Lawrence, R. Alvis, D. Olson, *Microsc. Microanal.* **14**(S2), 1004 (2008).

117. T. J. Prosa, D. Lawrence, D. Olson, D. J. Larson, E. A. Marquis, *Microsc. Microanal.* **15**(S2), 298 (2009).
118. K. Thompson, B. P. Gorman, D. J. Larson, B. van Leer, L. Hong, *Microsc. Microanal.* **12**(S2), 1736 (2006).
119. M. Ohring, *The Materials Science of Thin Films*, Academic Press, New York (1992).
120. L. A. Giannuzzi, F. Stevie, *Introduction to Focused Ion Beams, Instrumentation, Theory, Techniques and Practice*, Springer, New York (2005).
121. R. Doering, N. Nishi, *Handbook of Semiconductor Manufacturing Technology*, CRC Press, Boca Raton, FL (2008).
122. B. P. Gorman, D. Diercks, M. J. Kaufman, R. M. Ulfig, D. Lawrence, K. Thompson, D. J. Larson, *Microsc. Microanal.* **12**(S2), 1720 (2006).
123. B. P. Gorman, D. Diercks, *Microsc. Microanal.* **13**(S2), 822 (2007).
124. B. P. Gorman, *Microsc. Microanal.* **13**(S2), 1616 (2007).
125. P. R. Bevington, K. D. Robinson, *Data Reduction and Error Analysis for the Physical Sciences*, Third Ed., McGraw-Hill Higher Education, New York, NY (2003).
126. A. H. Cotrell, B. A. Bilby, *Proceedings of the Physical Society of London* **A62**, 49 (1949).
127. K. Thompson, P. L. Flaitz, P. Ronsheim, D. J. Larson, T. F. Kelly, *Science* **317**, 1370 (2007).
128. R. Mukai, S. Ozawa, H. Yagi, *Thin Solid Films* **270**, 567 (1995).
129. D. Mangelinck, J. Y. Dai, J. Pan, S. K. Lahiri, *Appl. Phys. Lett.* **75**, 1736 (1999).
130. D. Mangelinck, J. Y. Dai, S. K. Lahiri, C. S. Ho, T. Osipowicz, *Materials Research Society Symposium Proceedings* **564**, 163 (1999).
131. P. S. Lee, K. L. Pey, D. Mangelinck, J. Ding, A. S. T. Wee, L. Chan, *IEEE Electron Device Letters* **22**, 568 (2001).
132. P. Ronsheim, J. McMurray, P. Flaitz, C. Parks, K. Thompson, D. J. Larson, T. F. Kelly, *Proceedings of the American Institute of Physics* **931**, 129 (2007).
133. Y.-C. Kim, P. Adusumilli, L. J. Lauhon, D. N. Seidman, S.-Y. Jung, H.-D. Lee, R. L. Alvis, R. M. Ulfig, J. D. Olson, *Appl. Phys. Lett.* **91**, 113106 (2007).
134. O. Cojocaru-Miredin, D. Mangelinck, K. Hoummada, E. Cadel, D. Blavette, B. Deconihout, C. Perrin-Pellegrino, *Scripta Mater.* **57**, 373 (2007).
135. D. Mangelinck, K. Hoummada, A. Portavoce, C. Perrin, R. Daineche, M. Descoins, D. J. Larson, P. H. Clifton, *Scripta Materi.* **62**, 568 (2010).
136. P. Adusumilli, L. J. Lauhon, D. N. Seidman, C. E. Murray, O. Avayu, Y. Rosenwaks, *Appl. Phys. Lett.* **94**, 103113 (2009).
137. K. Hoummada, C. Perrin-Pellegrino, D. Mangelinck, *J. Appl. Phys.* **106**, 113525 (2009).
138. P. Adusumilli, C. E. Murray, L. J. Lauhon, O. Avayu, Y. Rosenwaks, D. N. Seidman, *Transactions of the Electrochemical Society* **19**, 303 (2009).
139. L. G. Harrison, *Transactions of the Faraday Society* **57**, 1191 (1961).
140. R. M. Ulfig, K. Thompson, R. L. Alvis, D. J. Larson, P. Ronsheim, *Microsc. Microanal.* **13**(S2), 828 (2007).

141. R. M. Ulfig, K. Thompson, R. L. Alvis, J. H. Bunton, B. P. Gorman, D. J. Larson, *Transactions of the Electrochemical Society* **6**, 14 (2007).

142. S. Mutas, C. Klein, S. S. A. Gerstl, *Ultramicrosc.* **111**, 546 (2011).

143. T. Kelwing, A. Naumann, M. Trentzsch, B. Trui, L. Hermann, S. Mutas, F. Graetsch, R. Carter, R. Stephan, P. Kucher, W. Hansch, *IEEE Electron Device Letters* **31**, 1149 (2010).

144. M. J. Galtrey, R. A. Oliver, M. J. Kappers, C. J. Humphreys, D. J. Stokes, P. H. Clifton, A. Cerezo, *Appl. Phys. Lett.* **90**, 061903 (2007).

145. B. P. Gorman, A. G. Norman, Y. Yan, *Microsc. Microanal.* **13**, 493 (2007).

146. M. J. Galtrey, R. A. Oliver, M. J. Kappers, C. McAleese, D. Zhu, C. J. Humphreys, P. H. Clifton, D. J. Larson, A. Cerezo, *Appl. Phys. Lett.* **92**, 041904 (2008).

147. M. J. Galtrey, R. A. Oliver, M. J. Kappers, C. J. Humphreys, P. H. Clifton, D. J. Larson, D. W. Saxey, A. Cerezo, *J. Appl. Phys.* **104**, 013524 (2008).

148. M. J. Galtrey, R. A. Oliver, M. J. Kappers, C. McAleese, D. Zhu, C. J. Humphreys, P. H. Clifton, D. J. Larson, A. Cerezo, *Phys. Stat. Sol. B* **245**, 861 (2008).

149. M. Müller, A. Cerezo, G. D. W. Smith, L. Chang, S. S. A. Gerstl, *Appl. Phys. Lett.* **92**, 233115 (2008).

150. M. Kodzuka, T. Ohkubo, K. Hono, F. Matsukura, H. Ohno, *Ultramicrosc.* **109**, 644 (2009).

151. D. Zhu, M. J. Kappers, C. McAleese, D. M. Graham, G. R. Chabrol, N. P. Hylton, P. Dawson, E. J. Thrush, C. J. Humphreys, *J. Crys. Growth* **298**, 504 (2007).

152. I. Farrer, M. J. Murphy, D. A. Ritchie, A. J. Shields, *J. Crys. Growth* **251**, 771 (2003).

153. T. Nakai, S. Iwasaki, K. Yamaguchi, *Jpn. J. Appl. Phys.* **43**, 2122 (2004).

154. M. Geller, C. Kapteyn, L. Müller-Kirsch, R. Heitz, D. Bimberg, *Appl. Phys. Lett.* **82**, 2706 (2003).

155. A. M. Beltrán, E. A. Marquis, A. G. Taboada, J. M. Garcia, J. M. Ripalda, S. I. Molina, *Ultramicrosc.*, **111**, 1073 (2011).

156. B. C. Lee, C. P. Voskoboynikov, C. P. Lee, *Phys. E* **24**, 87 (2004).

157. D. Alonso-Alvarez, B. Alén, J. M. García, J. M. Ripalda, *Appl. Phys. Lett.* **9**, 263103 (2007).

158. S. I. Molina, A. M. Beltrán, T. Ben, P. L. Galindo, E. Guerrero, A. G. Tabodada, J. M. Ripalda, M. F. Chisholm, *Appl. Phys. Lett.* **94**, 043114 (2009).

159. W. Lu, P. Xie, C. M. Lieber, *IEEE Trans. Electron Devices* **55**, 2859 (2008).

160. M. C. Putnam, M. A. Filler, B. M. Kayes, M. D. Kelzenberg, Y. Guan, N. S. Lewis, J. M. Eiler, H. A. Atwater, *Nano Lett.* **8**, 3109 (2008).

161. T. J. Prosa, R. Alvis, L. Tsakalakos, V. S. Smentkowski, *J. Microsc.* **239**, 92 (2010).

162. D. R. Diercks, B. P. Gorman, C. L. Cheung, G. Wang, *Microsc. Microanal.* **15**(S2), 254 (2009).

163. D. E. Perea, J. E. Allen, S. J. May, B. W. Wessels, D. N. Seidman, L. J. Lauhon, *Nano Lett.* **6**, 181 (2006).

164. D. E. Perea, J. L. Lensch, S. J. May, B. W. Wessels, L. J. Lauhon, *Appl. Phys. A* **85**, 271 (2006).
165. D. E. Perea, E. Wijaya, J. L. Lensch-Falk, E. R. Hemesath, L. J. Lauhon, *J. Solid State Chem.* **181**, 1645 (2008).
166. T. Xu, J. P. Nys, B. Grandidier, D. Stiévenard, Y. Coffinier, R. Boukherroub, R. Larde, E. Cadel, P. Pareige, *J. Vac. Sci. Technol. B* **26**, 1960 (2008).
167. D. E. Perea, E. R. Hemesath, E. J. Schwalbach, J. L. Lensch-Falk, P. W. Voorhees, L. J. Lauhon, *Nature Nanotech.* **4**, 315 (2009).
168. S. Zhang, E. R. Hamesath, D. E. Perea, E. Wijaya, J. L. Lensch-Falk, L. J. Lauhon, *Nano Lett.* **9**, 3268 (2009).
169. R. A. Schlitz, D. E. Perea, J. L. Lensch-Falk, E. R. Hemesath, L. J. Lauhon, *Appl. Phys. Lett.* **95**, 162101 (2009).
170. F. Li, T. Ohkubo, Y. M. Chen, M. Kodzuka, F. Ye, D. R. Ou, T. Mori, K. Hono, *Scripta Materialia* **63**, 332 (2010).
171. E. Talbot, R. Larde, F. Gourbilleau, C. Dufour, P. Pareige, *Euro. Phys. Lett.* **87**, 26004 (2009).
172. F. Iacona, C. Bongiorno, C. Spinella, S. Boninelli, F. Priolo, *J. Appl. Phys.* **95**, 3723 (2004).
173. J. Wang, X. F. Wang, Q. Li, A. Hryciw, A. Meldrum, *Philos. Mag.* **87**, 11 (2007).
174. J. S. Moore, K. S. Jones, H. Kennel, S. Corcoran, *Ultramicrosc.* **108**, 536 (2008).
175. K. Inoue, F. Yano, A. Nishida, T. Tsunomura, T. Toyama, Y. Nagai, M. Hasegawa, *Appl. Phys. Lett.* **92**, 103506 (2008).
176. K. Inoue, F. Yano, A. Nishida, H. Takamizawa, T. Tsunomura, Y. Nagai, M. Hasegawa, *Appl. Phys. Lett.* **95**, 043502 (2009).
177. K. Inoue, F. Yano, A. Nishida, H. Takamizawa, T. Tsunomura, Y. Nagai, M. Hasegawa, *Ultramicrosc.* **109**, 1479 (2009).
178. A. K. Kambham, J. Mody, M. Gilbert, S. Koelling, W. Vandervorst, *Ultramicrosc.* **111**, 535 (2011).
179. W. Vandervorst, M. Jurczak, J. Everaert, B. J. Pawlak, R. Duffy, J. Del-Agua-Bomiquel, T. Poon, *J. Vac. Sci. Technol. B* **26**, 396 (2008).

11

RAMAN SPECTROSCOPY OF CARBON NANOTUBES AND GRAPHENE MATERIALS AND DEVICES

Marcus Freitag and James C. Tsang
IBM T. J. Watson Research Center
Yorktown Heights, NY 10598, USA

11.1. INTRODUCTION

In recent years, there has been great interest in using either carbon
nanotubes or graphene in future integrated circuit (IC) technologies.
This has been spurred by doubts about whether the established path for
improvements in high performance silicon IC technology, the continuing
reduction in the dimensions of active devices with the accompanying
increase in their density, can be maintained using the traditional materials of silicon and silicon dioxide.[1] In recent years for example, the continued reduction of device dimensions including gate oxide thicknesses
has required the use of new gate oxides and gate materials, as well as
the traditional careful control of the geometry of the device.[1] As quasi-1
and 2 dimensional systems, devices built from carbon nanotubes and
graphene can achieve extremely high densities while maintaining large
carrier mobilities, qualities that are essential for future high performance
electronic devices. The stability of the bonding between the carbon
atoms in these materials with its similarity to the resonance bonding
in benzene[2] suggests that while these systems can have large surface
area to volume ratios, the properties of carbon nanotubes and graphene
could be relatively insensitive to chemical defects arising from reactions
with substrates under the channel or coatings on top of it, a major
advantage in the creation of a robust and reliable technology. In particular, the absence of intrinsic dangling bonds normal to the carbon planes
removes a major family of chemical perturbation to the electronic properties of the carbon structures. Of course the small dimensions of these
systems mean that electrostatic effects must still be carefully controlled
to prevent doping effects due to trapped charges in the substrate or any
covering layer and any other long range interactions with the substrate
and any overlayer.

For all their nominal advantages, any real efforts to use carbon nanotubes or graphene in practical electronic devices will require measurement tools that quantitatively characterize (1) these materials as they are synthesized or received to assure their quality with respect to critical parameters such as the diameter in the case of carbon nanotubes, deliberate or accidental doping, etc., (2) the devices based on these materials as they are being fabricated to verify the accuracy and quality of the fabrication process, and (3) the fabricated circuits as they switch and process information to monitor the presence of parasitic circuit elements and other problems that would degrade the performance. Such detailed monitoring of all stages of the creation of working circuits is needed to understand where faults and failures might occur in any new technology.[3]

Raman spectroscopy has been widely used to characterize both carbon nanotubes and graphene.[4,5] In this article, we review the information that can be derived from Raman scattering measurements of carbon nanotubes and graphene and devices fabricated from them. Given the small amount of material in either a carbon nanotube or a graphene layer working as the channel of a field-effect transistor (FET), questions can also be asked about the practicality of Raman spectroscopy as a characterization tool. Raman scattering from bulk materials is generally thought to produce small signals which require lengthy integration times to obtain useful signal-to-noise ratios. We therefore first show experimental results demonstrating that Raman spectra can be obtained from single carbon nanotubes and individual graphene samples with good signal-to-noise ratios. We describe the optical phonons that can be observed in these carbon based systems. An understanding of these excitations shows how Raman scattering can provide information about intrinsic and extrinsic doping effects, temperature, strain, the diameter of a carbon nanotube, the number of layers in multilayer graphene systems and other important physical parameters that play a major role in the transport properties of carbon based field effect transistors. We then review the physical mechanisms responsible for vibrational Raman signals in carbon nanotubes and graphene to show that the intensities of the experimental results are usable. Finally, it is shown that Raman spectroscopy can be used to monitor important physical parameters when these materials are in working devices.

11.2. RAMAN SCATTERING FROM CARBON NANOTUBES AND GRAPHENE

Raman scattering is the inelastic scattering of light by excitations of a scattering medium.[6,7] The energy of the scattered light is shifted with respect to the energy of light incident on the sample. The difference in the energies of the incident and scattered photon is the energy of the excitation scattering the incident light. In this article, the excitations will be phonons or the vibrations of the atoms making up the material of interest. This energy difference or shift and the symmetry properties of the matrix connecting the incident and scattered light and the scattered phonon, are characteristic of the scattering medium and can be used to identify the scatterer and describe its properties. Since the energy shifts are usually small compared to the energy of the incident light, and the Raman signals are orders of magnitude smaller than the intensity of the incident light, Raman scattering enjoyed a major renaissance with the discovery of the laser which generates extremely monochromatic light beams ideally suited for Raman scattering measurements. Over the last quarter century, Raman scattering has become a useful tool for characterizing modern Group IV, III-V and II-VI materials, devices and circuits, even as the devices and circuits have achieved dimensions smaller than the diffraction limit.[8]

The small size of the energy shifts in Raman scattering compared to the excitation energies used in the measurements makes Raman spectroscopy a simpler measurement than infrared spectroscopy where the spectral band that has to be measured is comparable to the spectral energy. The use of excitation wavelengths in the visible in Raman scattering experiments also gives Raman spectroscopy a substantial advantage in its spatial resolution over infrared spectroscopy which works at much longer wavelengths and has much poorer spatial resolution due to the diffraction limit. This is a significant handicap in looking at the local properties of micro and nanoscale devices.

11.2.1. Raman Scattering

The first order Raman scattering in a material such as graphene is schematically described by the diagrams in Fig. 11.1.[9] Detailed discussions of the physics behind the diagrams can be found in the first two volumes of Cardona and Guntherodt[6] and Loudon.[7] The individual diagrams in Fig. 11.1 are graphical representations of the terms that arise from a time dependent perturbation theory analysis of the probability of a Raman scattering event using Fermi's Golden Rule. The diagrams describe sequential processes such as the initial creation of an electron-hole pair (solid arrows) by the incident photon (zigzag lines), the scattering of either the electron or hole by a single phonon (dotted arrow), and the emission of the Raman shifted photon by the scattered electron-hole pair (Figs. 11.1(a) and (b)), the creation of a low energy electron-hole pair by the incident and emitted photons, and the subsequent emission of the phonon by the low energy electron-hole pair (Fig. 11.1(e)) and other possible terms including the direct decay of the incident photon into the scattered photon and the emitted phonon

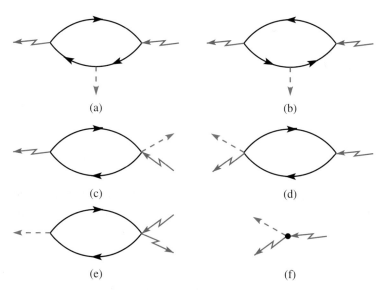

(a)   (b)

(c)   (d)

(e)   (f)

Fig. 11.1 Schematic representation of Raman scattering processes. The solid arrows represent electronic excitations, the dotted arrows phonons, and the zigzag lines the incident and scattered photons. Reproduced from D. M. Basko.[9]

(Fig. 1f). Higher order scattering processes involving the emission of more than one phonon are described by introducing additional electron-phonon vertices. Figure 11.1 incorporates the energy, $h\nu$, and wavevector k conservation relationships

$$h\nu_s = h\nu_i \pm \frac{h}{2\pi}\omega_o$$

$$k_s = k_i \pm k_o$$

where the subscripts refer to the scattered (s), and incident (i) photons while (o) represents the excitation of the scattering medium. When the scattered light is at a lower energy and longer wavelength, and an excitation of the sample is emitted in the scattering event, the process is called a Stokes process. When the scattered light is at a higher energy and shorter wavelength, so that the excitation is absorbed, it is an anti-Stokes process.

The wavevector conservation condition limits the Raman scattering in crystalline materials to long wavelength excitations since the wavelength of light is large compared to the interatomic spacing in any crystalline material. Wavevector conservation for Raman scattering from crystalline materials can be satisfied by the creation of a single long wavelength excitation, the generation of pairs of excitations with equal and opposite wavevectors, or multiple finite wavelength excitations whose total wavevector is zero. The presence of defects in a crystalline solid relaxes the wavevector conservation condition and can allow normally Raman inactive modes to be observed. The coupling of the incident and scattered photons in Fig. 11.1 by an excitation of the scatterer such as a phonon means that the tensor describing the Raman scattering process will reflect the symmetry of the scatterer.[6]

While the excitation involved in the Raman scattering event is a phonon, Fig. 11.1 shows that the Raman scattering process also involves the electronic system of the scatterer. The physical interactions involved in the Raman process are the electron-photon interaction and the electron-phonon interaction.[6] The non-resonant excitations of the electronic system in off-resonance Raman scattering are virtual excitations. Even when the photons are in resonance with the electronic states of the scattering material, the resonant Raman process requires that the electronic excited states are only scattered by the elementary excitation specified by ω_o and k_o. Figure 11.1(a) can be strongly resonant if there are electronic transitions of the scattering medium that are at the same energy as the incident and scattered photons. In fact, the intensity of the resonant Raman scattering process described by Fig. 11.1(a) is inversely proportional to some power of the lifetime of the electronic excited states.[6] This fact will be discussed later in this article as we consider the intensity of the Raman signals, especially in the case of carbon nanotubes. Critical points in the electronic structure can produce significant enhancements of the Raman scattering intensity due to resonant Raman processes and these will be important in satisfying the practical requirement that the signals from small samples are detectable in reasonable measurement times. Resonances between the incident and scattered light and the critical points in the electronic structure produce large enhancements in the magnitude of the Raman susceptibility.

11.2.2.   Raman Spectra of Planar Carbon Systems

Structurally, the three carbon systems are closely related. Graphene is a sp^2-bonded honeycomb lattice of carbon atoms. It's Bravais lattice is a hexagonal (equilateral triangular) lattice with an identical two-atom basis (carbon atoms A and B). Graphite consists of graphene sheets stacked in an AB sequence, where the A' atoms in every even layer sit between the B atoms of odd layers, while the B' atoms in every even layer sit between the centers of the hexagons defined by the odd layers (Fig. 11.2). Carbon nanotubes can be derived from graphene through

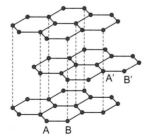

Fig. 11.2   AB stacking order of graphite. The two carbon atoms in the graphene unit cell, A and B, are depicted in red and blue.

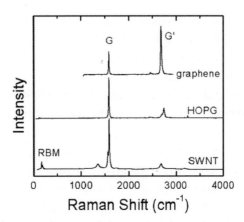

Fig. 11.3 Raman spectra of graphene, graphite, and carbon nanotubes. Prominent features include the G-line near $1600\,\text{cm}^{-1}$, G' or 2D line near $2700\,\text{cm}^{-1}$, D-line near $1350\,\text{cm}^{-1}$, and the radial breathing mode, RBM (in the case of carbon nanotubes) between 100 and $400\,\text{cm}^{-1}$, dependent on diameter. Reproduced with permission of M. S. Dresselhaus *et al.*[5]

the cylindrical folding of finite width strips of graphene into tubes.[10] All three systems have identical hexagonal bonding and very similar nearest and next-nearest neighbor configurations.

The Raman spectra of graphene, graphite, and a carbon nanotube are shown in Fig. 11.3 for energy shifts between 100 and $4000\,\text{cm}^{-1}$.[5] The energies and displacement patterns of the Raman active long wavelength optical phonons are largely determined by the nearest and next-nearest neighbor interactions and are similar for graphene and carbon nanotubes. In so far as interlayer interactions are weak and negligible, graphite is just a stack of graphene layers. Similarly, if the curvature around a carbon nanotube is small i.e. the diameter of the tube is large compared to the size of the hexagonal lattice, the energies of the optical phonons of a single wall carbon nanotube will be close to those of graphene. The long wavelength acoustic phonons of graphene and carbon nanotubes can be quite different given their strong dispersion and the periodic boundary conditions that apply to carbon nanotubes. The electronic structure of carbon nanotubes consists of one-dimensional sub-bands for similar reasons, which affects the Raman intensities. For small diameter carbon nanotubes, where the graphene planes are strongly curved, significant differences between the carbon nanotube and graphene properties are expected.

Given the structural similarities, a number of the features of the Raman spectra of graphite, graphene and carbon nanotubes are similar. The sharp lines and bands in Fig. 11.3 can be separated into three different groups. All three systems show a very strong sharp line around $1585\,\text{cm}^{-1}$ which historically has been called the G line. They also show broader structures between 2600 and $3200\,\text{cm}^{-1}$. These have been called G' or 2D line for the band around $2700\,\text{cm}^{-1}$, and 2G for the structure near $3200\,\text{cm}^{-1}$. In the presence of disorder, all three systems in Fig. 11.3 show another band at about $1350\,\text{cm}^{-1}$ or half the energy of the 2D band. Graphite and single wall carbon nanotubes show lower lying sharp lines at Raman shifts below $400\,\text{cm}^{-1}$. For graphite, there is a sharp line at $42\,\text{cm}^{-1}$ (not shown in Fig. 11.3) which is the rigid layer mode, and involves the shear motion of neighboring graphite layers. For the carbon nanotubes, the low lying Raman line arises from a breathing motion of the wall of the nanotube and it is called the RBM or Radial Breathing Mode.[4] Its energy depends on the diameter of the tube, and this Raman line is widely used to help determine the diameters of individual tubes. The results in Fig. 11.3 show that Raman spectra can be obtained using conventional instrumentation from a single layer of graphene and a single carbon nanotube.

Figure 11.4 shows how the electronic structure of graphene, and the closely related graphite and carbon nanotube systems can support different types of resonant Raman processes, in particular for both the one phonon G line and the 2 phonon 2D/G' band in Fig. 11.3. The band structures show the linear dispersion around the Dirac point that is characteristic of graphene. The top half of the figure shows a first order Raman process involving the excitation of one long wavelength phonon where only the incident or scattered photon but not both can be in resonance with an electronic transition of the graphene. Alternatively, a single phonon with an arbitrary wavevector can scatter the excited electron to another real state in the electronic structure but in general the real state shown to the right of the top of Fig. 11.4 will not correspond to a resonant emission of the scattered photon. The bottom half shows a two phonon scattering process where both the incident and the scattered photons are in resonance with the electronic excitations of the graphene. The latter situation has been rare in most materials[54] but will

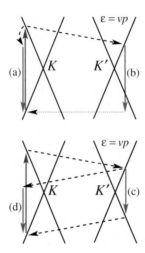

Fig. 11.4   Single and double resonance processes in graphene. (a,b) Single-resonant process, where the incoming photon is resonant, but the scattered photon is non-resonant.
(c,d) Double-resonant process, where both the incoming and scattered photons are resonant.
The solid lines represent the low-energy electronic dispersion of graphene (energy vertical,
momentum horizontal). Solid arrows represent electronic transitions due to the incoming and
scattered photons. Dashed arrows indicate scattering of electrons or holes by phonons. The
horizontal dotted arrow represents impurity scattering. Reproduced from D. M. Basko.[55]

be seen to be common in the graphene based systems. Such double resonances are a distinctive feature of the graphene based systems and have
been invoked to explain several unusual features of the Raman spectra
of these materials. These include the strength of the 2D/G' Raman scattering, the strong dependence of the intensity of the Raman spectrum of
a carbon nanotube on the excitation wavelength, and the dependence of
the Raman energy on excitation wavelength for the D and 2D/G' lines.

The phonon dispersion for graphite is shown in Fig. 11.5, as derived
from an experimental determination by X-ray energy loss measurements
and computationally by fitting the experimental data by a fifth nearest
neighbor force constant model.[11] Two vibrational modes of particular
interest to Raman scattering are the highest energy phonons at $k = 0$ (Γ-
point) with an energy near 196 meV or 1580 cm^{-1} and the zone boundary
K point with an energy about 165 meV or 1350 cm^{-1}. The strong sharp
lines near 1585 cm^{-1} in Fig. 11.3 (G-lines) for all three carbon materials
are due to scattering of the incident photon by the highest energy $k = 0$
optical phonon in Fig. 11.5. In carbon nanotubes, the G-line splits into

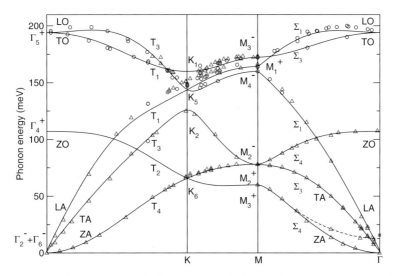

Fig. 11.5   Experimentally determined and calculated phonon dispersion curves of graphite. Graphene shows similar phonon dispersion. The Raman G-band is due to the highest-energy optical phonons at k = 0 (Γ-point). The G' or 2D-band is due to two optical phonons near k = K with energy near $1350\,\mathrm{cm}^{-1}$. Reproduced from M. Mohr *et al.*[11]

longitudinal optical (LO) and transverse optical (TO) components due to curvature. It has also been shown that the broader line near $2700\,\mathrm{cm}^{-1}$ in Fig. 11.3 arises from the excitation of two K point optical phonons with equal and opposite crystal momenta or k values. The weak structure near $3200\,\mathrm{cm}^{-1}$ is due to the excitation of pairs of long wavelength phonons with wavevectors close to the $k = 0$ phonon responsible for the strong $1585\,\mathrm{cm}^{-1}$ mentioned earlier. Careful examination of the Raman spectra in Fig. 11.3 also shows a weak band near $1350\,\mathrm{cm}^{-1}$. This arises from the excitation of a single K point highest energy optical phonon, and is the fundamental to the two phonon $2700\,\mathrm{cm}^{-1}$ line mentioned previously. The displacement patterns for these two phonons are shown in Fig. 11.6.

The zone boundary K point mode is the symmetric breathing mode of the benzene ring (A_1) that is the basis of the three different carbon systems. This mode is the characteristic first order Raman mode of benzene. It appears only weakly in Raman scattering from well ordered graphene and its derivatives because the displacement pattern does not correspond to a long wavelength excitation of graphene. The highest energy zone center optical phonon mode which produces the G line shows

Γ vibrations:

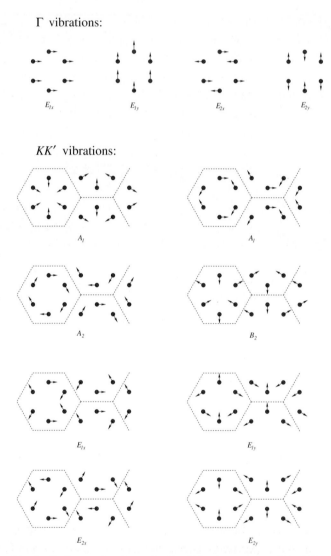

KK' vibrations:

Fig. 11.6 Atomic displacement patterns for the highest energy k = 0 and K point optical phonons of graphene. Reproduced from D. M. Basko.[55]

the same local atomic displacement pattern for all three materials (E_{2g}) and is given at the top of Fig. 11.6. The similarities of the G line in the three different carbon materials shows that the energy of this mode is only weakly perturbed by the different long range environments of the three materials.

The strong Raman bands around $2700\,\mathrm{cm}^{-1}$ in Fig. 11.3 arise from the resonant excitation of the two highest energy optical phonons at the K point of the graphite Brillouin zone, with equal and opposite momenta as shown in Figs. 11.4(c)–(d).[12] The result of such an excitation is a net long wavelength excitation to which the long wavelength incident and scattering photons can couple. The intensity of this scattering can show significant changes as we go from graphite to graphene to a carbon nanotube, and with doping in graphene, and has been the subject of considerable study in recent years. These changes are attributed to the dependence of the Raman scattering intensities on resonances in the electronic structure of graphite, graphene and carbon nanotubes and have been quantitatively explained in the context of the diagram in Figs. 11.4(c)–(d) and the electronic structure of these materials.[12] Unlike most Raman lines which show no dispersion, these Raman lines are dispersive, with a linear dependence of the measured Raman shift on excitation wavelength of the order of $100\,\mathrm{cm}^{-1}/\mathrm{eV}$. As mentioned earlier, the $2700\,\mathrm{cm}^{-1}$ band is the second harmonic of the defect induced D line at $1350\,\mathrm{cm}^{-1}$. The D line also shows dispersion, only at half the rate of the 2D line, consistent with its identification as the fundamental of the $2700\,\mathrm{cm}^{-1}$ band. The dispersive character of these lines can be seen from Figs. 11.4(c)–(d) where changes in the excitation energy will result in changes in the wavevectors of the scattered phonons.

11.3. PERTURBATIONS OF THE RAMAN PHONON FREQUENCIES

Raman spectroscopy has been used to characterize many aspects of silicon and silicon integrated circuits including for example local stresses. Such Raman measurements take advantage of the spatial resolution of the optical technique in being able to spatially resolve the properties of devices on the sub-micron scale.[8,13] Parameters that are important to the behavior of Si circuits include doping, the temperature of operation of the devices and circuits, the presence of strains due to geometric effects and processing, and also disorder which can be introduced by various processing steps. We now consider how doping, temperature, strains, and disorder change the Raman active phonons on the carbon based systems.

11.3.1. Doping Effects

The highest energy optical phonons at $k = 0$, and $k = K$ shown in Figs. 11.5 and 11.6 for graphite, graphene, and carbon nanotubes which generate the G, D and 2D/G′ lines in Fig. 11.3 interact strongly with carriers.[14−20] Detailed calculations of the dispersion curves of these phonons show an anomalous linear dependence of the energy as a function of wavevector near the high symmetry points at $k = 0$ and K.[14] These are due to Kohn anomalies where the high density of states of carriers in these low dimensional systems produces observable screening of these high symmetry phonons. This screening is very sensitive to the position of the Fermi level in graphene and carbon nanotubes. When the Fermi level is at the Dirac point of graphene, both the Γ and K point phonons can excite electron hole pairs across the Fermi level.[16] If the graphene layer is part of an FET, application of a gate voltage can move the Fermi level so that either the initial states for optical phonon excited electronic transitions are empty or the final states filled. This means that the phonons can no longer excite electron-hole pairs across the Fermi level. This generates both the change in the linewidth of the G band in graphene with applied gate voltage and Fermi level shown in Fig. 11.7, as well as the shift in the energy of the G line.[16,18] A schematic illustration of how shifts in the Fermi level near the Dirac point of graphene change the excitations of the electronic system degenerate with the optical phonon energies as described above is shown on the right. Similar behavior has been observed in metallic carbon nanotubes, where filling of electronic states either allows or blocks phonon excited interband transitions (see Sec. 11.7.3).[19,20]

11.3.2. Temperature

We showed in Fig. 11.7 that the energies of several of the characteristic Raman active phonons of graphene depend on the carrier density. Like most materials, the physical properties of graphene and its related

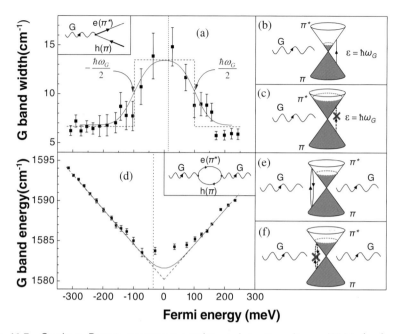

Fig. 11.7   Graphene Raman spectroscopy under varying gate voltages. Width (a–c) and position (d–f) of the G-band in electrostatically-gated graphene. At low doping the G-band phonon can decay into electron-hole pairs (b), while at high doping (n- or p-type), this decay path is forbidden due to the Pauli Exclusion Principle (c). The G-band width is therefore reduced at high doping (a). In addition, the phonon energy is renormalized due to the inter-action with virtual electronic excitations (e). The renormalization is reduced at higher doping (f), giving rise to a blue-shift in G-band energy (d). Reproduced from J. Yan *et al.*[16]

materials depend on the temperature. Anharmonic lattice effects which produce the change in the lattice constants of materials with changing temperature also change the phonon energies.[21] In the case of materials such as graphene and carbon nanotubes, the traditional anharmonic lattice effects involving the decay of an individual phonon into multiple phonons are supplemented by the decay of a phonon into electron-hole pairs and other related renormalizations. The temperature dependent changes in the optical phonon frequencies have been measured and are shown in Fig. 11.8 for G lines in single-layer graphene (SLG), bilayer graphene (BLG), graphite, and carbon nanotubes.[22,23] The measured shifts have been shown to be in good agreement with theoretical con-siderations including both the normal phonon-phonon interactions and

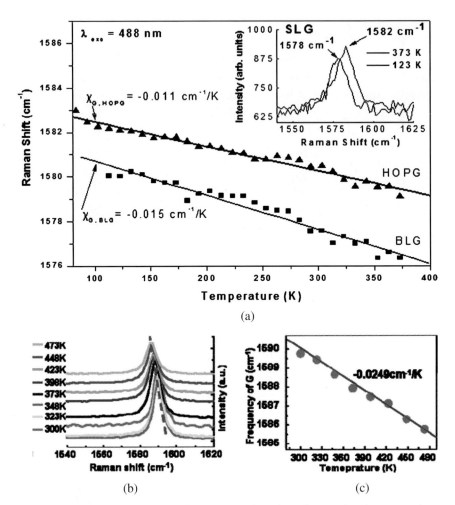

Fig. 11.8 Temperature-dependent Raman scattering of graphene and carbon nanotubes. Temperature dependence of the energy of the G mode of (a) SLG, BLG, graphite, and (b–e) a carbon nanotube. (a) Reproduced with permission of I. Calizo et al.[22] (b–c) Reproduced from Y. Zhang et al.[23]

also the electron-phonon interaction. We show later that Raman spectroscopy also provides a number of other experimental ways to measure the temperature of a sample[24] and that these alternative methods, combined with the temperature dependence of the phonon frequencies provide a very powerful methodology for the measurement of temperatures in carbon based systems.

11.3.3. Strain

Application of pressure to most materials produces changes in the frequencies of their phonons. Pressure reduces the lattice constants which produces a hardening of the force constants between atoms and increases the phonon energies. Application of uniaxial strains to materials also produces similar changes in the phonon energies. Symmetry reducing strains produce both splittings of normally symmetry degenerate phonon modes, and shifts in the phonon frequencies. Such measurements have been made in the carbon systems and results for graphene are shown in Fig. 11.9.[25,26] The experimental results are in good agreement with theoretical expectations. Strain plays a major role in determining the properties of Si based electronic devices and the ability to measure strains on a microscopic scale will be important for carbon based electronics given the small sizes of the carbon regions compared to the substrates and overlayers.

11.3.4. Disorder

Symmetry forbidden Raman scattering can be activated by disorder. Figure 11.10 shows how the Raman spectrum of graphite changes when it is ion implanted.[27] A light implanted ion (Be) produces a weak line at $1355 \, cm^{-1}$ which grows in strength as the ion mass increases. The Raman spectrum after implantation of phosphorus is identical to that of amorphous carbon. Figure 11.10 shows that disorder can activate the highest lying K point optical phonon, and that its intensity can be equal to that of the G line.

The D- and G-band intensity ratio I_D/I_G can also be associated with the graphitic crystallite size L_a. Tuinstra and Koenig[28] who experimented in 1970 on various samples of graphitic carbon, found an inverse relationship between I_D/I_G and L_a. More recently, Cancado et al.[29] generalized that result for arbitrary laser wavelength λ_L (Fig. 11.11): $L_a(nm) = 2.4 \times 10^{-10} \lambda_L^4 (I_D/I_G)^{-1}$.

Disorder effects have also been observed in the 2D-peak width in graphene. This is shown in Fig. 11.12 for epitaxial graphene grown on SiC substrates. Topographical studies of these samples showed regions

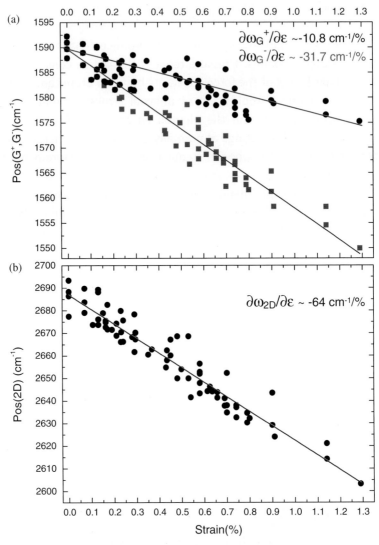

Fig. 11.9 Raman scattering of graphene under uniaxial strain. Strain dependence of (a) the graphene G line position, and (b) the graphene 2D-line position. Reproduced from T. M. G. Mohiuddin *et al.*[25]

of graphene with small grains of variable thickness, which had degraded Hall carrier mobilities. These regions were correlated with increases in the 2D spectral widths and also changes in the relative intensities of the 2D and G bands.[30] High-mobility samples were uniform both in

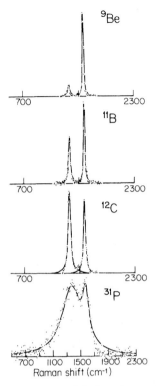

Fig. 11.10   Ion implantation, disorder, and D lines in graphite. G-line of graphite, ion implanted with successively heavier ions. Reproduced from B. S. Elman *et al.*[27]

topography measurements and Raman measurements, and the 2D width was below $40\,\mathrm{cm}^{-1}$.

11.4.   MEASUREMENTS OF INTENSITIES OF SYMMETRY ALLOWED RAMAN LINES

The Raman spectra in Fig. 11.3 are characterized by both the frequencies of the Raman lines, and also their intensities. The intensities are critical to any application of Raman spectroscopy for the characterization of devices and circuits since if the signals are too small, they could be undetectable against background signals arising from substrates and overlayers, or may require too long a measurement time to be practical. The intensity of Raman scattering is parameterized in terms of

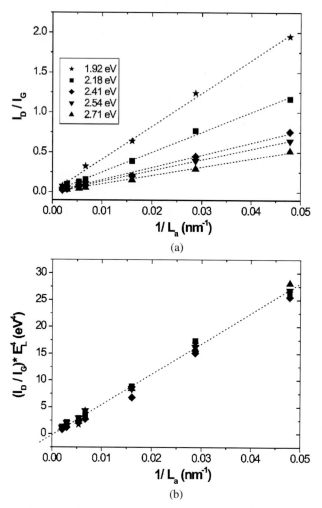

Fig. 11.11 Graphite crystallite size extracted from D over G intensity ratio. (a) D over G intensity ratio as a function of the inverse crystallite size for different laser energies. (b) The same data from above, but scaled with the fourth power of the laser energy. The data collapses on a single line. Reproduced from L. G. Cancado et al.[29]

either a scattering efficiency or a scattering cross section. The scattering efficiency for the long wavelength Raman active optical phonon of bulk Si under 1.9 eV (red) excitation is 1.68×10^{-3} sr^{-1} m^{-1} where the normalizations are for the solid angle of the scattered light that is collected in the measurement, and the optical thickness of the sample.[6] Because

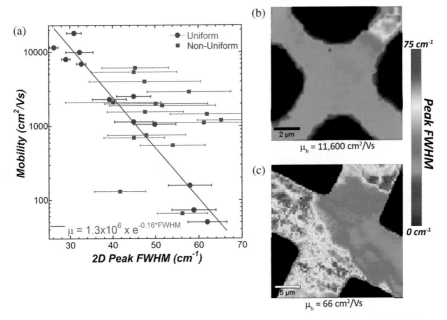

Fig. 11.12   Correlation of 2D peak width with graphene mobility. (a) 2D peak FWHM as a function of mobility. (b) Example of a uniform Hall cross. (c) Example of a Hall cross with non-uniform Raman signal. The trend shown in (a) is only valid for uniform samples. Reproduced from J. Robinson *et al.*[30]

of the large index of refraction of light in Si, this scattering efficiency for a 0.1 sr internal collection angle, and a 1 nm thick layer of Si would be 1.68×10^{-13}. Given the finite efficiency of monochromators and detectors, the overall efficiency for detection of Raman scattered photons is of the order of 10^{-15}. In strongly light absorbing materials such as Si, and the opaque carbon based materials, there are limits on how much light can be focused onto a sample surface without appreciable heating, and possible damage to the sample. Typically, for spot sizes of the order of 1 μm in diameter, the maximum useful focused power is less than a milliwatt which means that the rate of scattered photons from 1 nm of Si under 1.9 eV excitation is of the order of $1 s^{-1}$. For even state of the art detectors such as cooled CCDs, lengthy integration times would be needed to obtain experimentally usable spectra with reasonable signal to noise ratios in the absence of any background signals and dark counts arising from the detector. The Raman scattering efficiency for diamond is about 4% of that of Si so that Raman scattering from a 1 nm layer of diamond would be extremely difficult to detect with detectors whose dark count is of the order a few counts or less. Interestingly, the Raman efficiency of Ge under 1.9 eV excitation is almost 250× that of Si.[6] This means that a 1 nm thick layer of Ge would produce signals at the hundreds of detected photons/sec level which can readily be observed by modern optical detectors. The large magnitude of the Raman efficiency of Ge at 1.9 eV is due to the match between the 1.9 eV excitation energy and the E_1 gap of Ge[6] and points to how resonance Raman effects can greatly increase the intensity of Raman scattering, and the sensitivity of Raman scattering in studying small systems with limited amount of material.

The comparison of the Raman scattering efficiencies of the Group IV materials diamond, Si, and Ge for different excitation energies shows that the strength of Raman scattering depends strongly on the electronic structure of the sample and the match of the excitation energy to critical points in the electronic structure of the material. The Raman scattering efficiency of graphite has been measured to be 8×10^{-3} sr^{-1} m^{-1} for 2.41 eV (green) excitation or about 5× larger than that of Si in the red. This increase over that of Si means that graphite and graphene signal levels would be of the order of several counts per second, which is accessible to state of the art detectors though would dictate long measurement times. Since the Raman scattering efficiency of graphite increases strongly with increasing excitation energy above 2.5 eV, this situation will improve if higher energy photons are used.[31]

The electronic structures of graphite and graphene for excitation energies in the visible are quite similar, so that the G-line Raman scattering efficiencies per graphene layer can be expected to be similar as well. This need not be the case for carbon nanotubes. The optical properties of carbon nanotubes are dominated by the low dimensional van Hove singularities introduced by their quasi-1D character.[4] The optical properties, including the Raman scattering efficiencies, of carbon nanotubes at excitation energies in resonance with these van Hove singularities show strong enhancements in magnitude. These enhancements in intensity compensate for the small amount of material being studied in a single carbon nanotube experiment and produce readily observable Raman signals from a single carbon nanotube as shown in Fig. 11.3. Of course, for excitation energies that are away from the van Hove singularities of the carbon nanotube, the Raman scattering efficiency is very low, and Raman signals hard to detect.

11.4.1. Interference Enhancements of Raman Scattering Intensities

The sensitivity of Raman spectroscopy as a tool for the characterization of thin layer systems is improved by the use of substrates which produce multiple reflection effects and interference enhancements of Raman scattering intensities.[32] The physics behind these enhancements is the same as that underlying anti-reflection coatings. Nemanich *et al.*[32] introduced the concept of Interference Enhanced Raman Scattering, and the conditions required for such enhancements are identical to those invoked by Novoselov[33] and analyzed by many including Casiraghi *et al.* for enhancing the contrast in images of graphene layers on substrates.[34] The use of a spacer layer of appropriate thickness and dielectric constant between a highly reflecting substrate and a thin absorbing layer on top of the spacer produces an antireflection structure where the optical absorption at selected wavelengths is strongly enhanced by the thin absorbing layer due to the multiple reflections of the light in the multilayer structure. The critical parameter for this behavior is the thickness of the spacer layer which has to obey the well known quarter wavelength condition, $(2m + 1)\lambda/4$ where m is an integer and λ the wavelength of light in the spacer. Detailed calculations have shown that multilayer structures obeying the quarter wavelength condition produce significant enhancements of the optical contrast of graphene layers on Si-SiO$_2$ structures[34] and also strong enhancements of the Raman scattering intensities from the graphene overlayer even though the Si substrate is far from an ideal reflector. Such effects produce the results shown in Figs. 11.13 and 11.14.[35,36]

In Fig. 11.13, different thickness of graphite between bulk graphite and graphene were placed in a Si-SiO$_2$ stack with the oxide thickness of 300 nm. The dependence of the intensity of the Raman spectra on

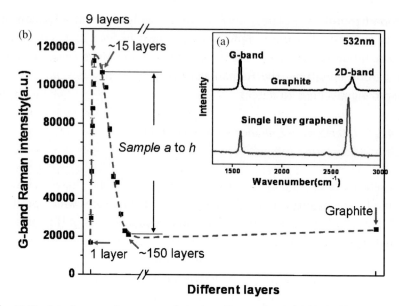

Fig. 11.13   Interference enhancement of graphene Raman signal. (a) Typical Raman spectra of graphite and graphene. (b) G-band Raman intensity as a function of layer number for graphene on Si/SiO$_2$ with 300 nm SiO$_2$. Reproduced from Y. Y. Wang et al.[35]

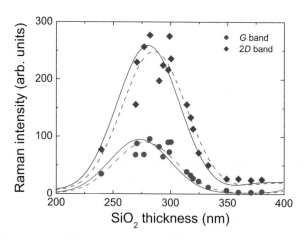

Fig. 11.14   Dependence of the intensity of the G line of graphene on the thickness of the SiO$_2$ spacer in a Si/SiO$_2$/graphene stack for a laser wavelength of 514.5 nm. The slight shift in profiles is due to the different energies of the G- and 2D transitions. Reproduced from D. Yoon et al.[36]

different thickness of graphite ranging from the bulk which is opaque and where there is no reflected light from the substrate, to graphene where only a few percent of the incident light is absorbed in the graphene for a single pass, and reflected light from the substrate can produce a significant enhancement of the optical fields at the graphene layer, shows a peak for graphene thicknesses where comparables fractions of the light are either reflected, or transmitted by the graphene layer. For such multilayer graphene thicknesses, there is a substantial interference enhancement of the Raman signal from the multilayer graphene and its signal is larger than that obtained from bulk graphite, a condition first discussed by Nemanich.[32] Figure 11.13 shows that the Raman signals from a single layer is comparable to that obtained from the bulk, even though there is much more material in the bulk layer.

The dependence of the graphene Raman signal on the SiO_2 spacer layer thickness is shown in Fig. 11.14.[36] The Raman intensities of the G and 2D bands are greatest for spacer thicknesses and excitation and emission wavelengths where the quarter wavelength condition is satisfied. As mentioned earlier, this is the condition that maximizes the optical contrast of the graphene layer since it maximizes the sensitivity of the optical response to small changes in the graphene overlayer. In the case of Raman scattering, the quarter wave condition produces an enhancement of the optical fields in the graphene layer, as well an interference enhancement of the Raman scattered light when both the directly emitted scattered light and the scattered light reflected from the substrate are added together. The last is in contrast to the normal case of the antireflection coating where the interference of the reflected light from the top layer and the reflected light from the substrate produce a cancellation of the reflected beam. The small displacement between the maxima for G and 2D as a function of oxide thickness in Fig. 11.14 is due to the fact that the wavelength of the 2D signal is different from that of the G signal. The interference conditions can therefore produce large enhancements of the Raman signal from graphene when there is constructive interference. Figures 11.13 and 11.14 show that the Raman intensities of graphene layers can be substantial and Raman spectra readily observed in commonly used substrate geometries, in particular, those pioneered by Geim *et al.*[33] if the correct excitation wavelength is used.

11.5.   INTERFACE AND RESONANT RAMAN EFFECTS

It was shown earlier that the Raman spectra of carbon based materials
such as graphite, graphene and carbon nanotubes show many qualitative
similarities but feature significant quantitative differences which could be
used to characterize these materials and devices and circuits fabricated
from them. A further example of this can be seen in Fig. 11.15. Graphene
was laid down on a Si-SiO$_2$ substrate with a hole cut in it. Therefore,
part of the graphene was supported by the oxide substrate and part of
the graphene layer was suspended over the hole. Such structures have
been of considerable interest recently since it has been shown that the
free standing suspended graphene has very high mobilities as compared
to the supported graphene which is in contact with a substrate.[38,39]

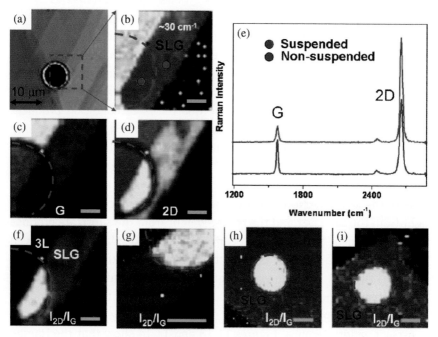

Fig. 11.15   Effect of the substrate on graphene Raman spectra. (a) Optical image of
graphene covering a hole in the substrate. (b) 2D-band FWHM used to differentiate between
single-layer and triple-layer graphene. (c) G-band intensity. (d) 2D-band intensity. (e) Raman
spectra taken at the spots indicated in (b). (f) 2D over G ratio. (g–i) 2D/G ratio for three
additional samples. Reproduced from Z. H. Ni et al.[37]

The spectra shown in Fig. 11.15 differ between the region supported by the oxide and that suspended over the hole. This has been attributed to localized doping of the supported graphene as compared to the suspended graphene, consistent with results where carrier doping changed the ratio of the G and 2D lines.[37] The increase in the strength of the 2D/G′ scattering for the intrinsic graphene should be observed for future high quality, high mobility graphene devices, even when they are not suspended.

It was mentioned in the discussion of Fig. 11.1 that while the frequency shifts seen in the Raman spectra of SP^2-hybridized carbon materials arise from the excitation of phonons, in particular the highest energy $k = 0$ and $k = K$ point optical phonons, the non-resonant Raman process involves virtual excitations of the electronic states where the energies of the excited states of the scattering system do not match the energies of the incident and scattered light. In the case of resonant Raman scattering as shown in Fig. 11.4, these energies do match so that the denominators of the perturbation theory expressions summarized in Fig. 11.1 are determined not by the difference in the energies of the light and the excitations of the scattering system but by the lifetimes of the excited states. Figure 11.16 shows how the Raman cross section of graphite increases rapidly with increasing excitation energy.[31] This is partly due to a general $(h\nu)^3$ dependence of the Raman scattering intensity with laser energy, and partly due to a broad π Plasmon resonance in the dielectric response of graphitic materials around 3.5 eV that becomes important above 2.5 eV since these transitions can contribute to the Raman susceptibility χ_{ph}. The entire Raman scattering rate of the Stokes process can be written as[24]:

$$R_{Stokes} = (n + 1)(E_L - \hbar\omega)^3 (\chi_{ph}(E_L))^2$$

where n is the phonon occupation number, and E_L and $\hbar\omega$ are laser and phonon energies. The use of higher energy photons is an alternative means of increasing the sensitivity of Raman scattering.

While the electronic structure of graphene and graphite, and their dielectric response are quite similar for photon excitations above 1 eV with no sharp structures, the electronic structure of carbon nanotubes shows strong additional structure in this energy range. This is due to the

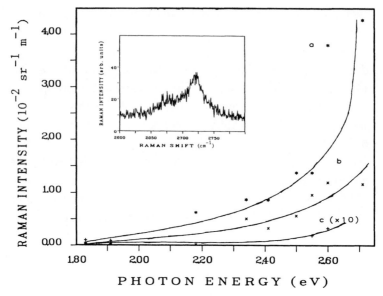

Fig. 11.16   Resonant Raman profiles of graphite. (a) G-band intensity. (b) 2D-band intensity. (c) Interlayer mode intensity at 42 cm^{-1}. Reproduced from K. Sinha et al.[31]

fact that while a carbon nanotube along its long axis has the same translational symmetry as graphene and graphite, it has a different symmetry around its circumference. In fact, the electronic structure of a carbon nanotube shares many of the features of a collection of one dimensional electronic bands. There are families of one dimensional critical points which introduce strong new features into the optical density of states. These additional features in the optical density of states contribute substantially to the Raman susceptibility discussed earlier.

Figure 11.17 shows a series of color-coded Raman spectra of an ensemble of carbon nanotubes acquired under varying laser excitation energies. The frequency of the radial breathing mode (RBM) of a carbon nanotube, ω_{RBM}, is inversely proportional to the nanotube diameter d,[40,41] and can therefore be used for identification purposes. There are slightly different proportionality factors and corrections to the $1/d$ behavior given in the literature. A. Jorio et al.[41] use $\omega_{RBM} \left(cm^{-1}\right) = 248/d$ (nm), while C. Fantini et al.[40] find $\omega_{RBM} \left(cm^{-1}\right) = 223/d$ (nm)+ 10. These values depend slightly on the environment of the nanotube. While it is possible to determine the diameter from the RBM energy

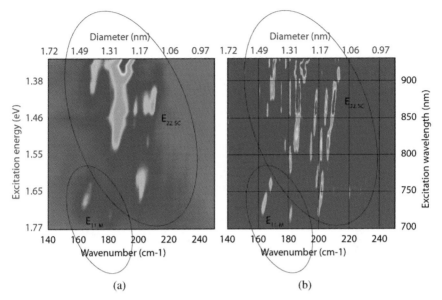

Fig. 11.17 Resonant Raman scattering of the radial breathing mode (RBM) of carbon nanotube buckypaper. (a) False-color plot of the Raman intensity as a function of Raman shift and excitation energy. (b) False-color plot of the second derivative with respect to wavenumber of the data in (a), showing the individual nanotube chiralities clearly resolved. Nanotubes were grown by laser-ablation. Diameters were assigned using $\omega_{RBM}(\text{cm}^{-1}) = 223/d\,(\text{nm}) + 10$.[40] The laser excitation is resonant with the E_{22} level in semiconducting nanotubes and the E_{11} level in metallic nanotubes (indicated).

alone, there are too many possible nanotube chiralities in a certain diameter range to make an unambiguous identification. However, resonant Raman spectroscopy enhances the signals of only those carbon nanotubes whose electronic levels are in resonance with the incoming or scattered photons. These electronic resonances can be quite distinct even for nanotubes with similar diameters. By using both, the RBM energy and the position of the electronic resonance, it is possible to resolve individual nanotube chiralities in ensembles and to make un-ambiguous chirality identifications.[40] Most of the RBM peaks in Fig. 11.17 are due to resonant Raman scattering involving the second sub-band (E_{22}) in different semiconducting carbon nanotube chiralities. The first sub-band in metallic carbon nanotubes contributes the peaks in the bottom left segment. Plots like these have been used extensively to determine the diameter and chirality distribution of carbon nanotubes grown under various methods, or nanotubes undergoing purification processes. In contrast to fluorescence excitation maps,[42] where the fluorescence at E_{11} and the absorption at E_{22} are used to make the identification, metallic carbon nanotubes are detected by the Raman method as well.

It was shown in Fig. 11.15 that there are substantial differences in the Raman spectra of graphene when it is supported on a substrate, and when it is suspended over a hole. Figure 11.18 shows resonant Raman profiles of a long carbon nanotube mainly supported by a silicon dioxide substrate, but also suspended over trenches in the oxide.[43] The peaks in the Raman scattering intensities near 2.2 and 2.4 eV arise

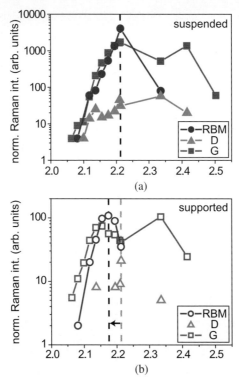

Fig. 11.18   Resonant Raman profiles of RBM, D, and G line of a carbon nanotube stretching across a trench in the Si/SiO$_2$ substrate. (a) RRS profiles for the suspended nanotube section. (b) RRS profiles for the supported nanotube sections, showing an overall decrease in intensity and shift to lower energies. A resonance of the incoming photon with an electronic state at about 2.2 eV (E_{33}) is responsible for the increased Raman intensities in (a) and (b). In the case of G and D-lines, the resonance with the scattered phonon at higher energy is also resolved. Reproduced from M. Steiner et al.[43]

from resonances of the incident and scattered photons with the E_{33} van Hove singularity of the carbon nanotube. Note that the energy of the RBM is small (~ 20 meV) compared to the width of the electronic transition, so that the resonances with the incident and scattered photons overlap, while in the case of the higher energy G and D (200 meV and 160 meV) they are clearly resolved. The peak resonant Raman intensities are greatest when the nanotube is suspended over a trench. The maximum in the resonant Raman profile of the supported regions is weaker than that of the suspended regions while the spectral widths of the resonances are broader for the supported regions. The electronic resonance is also shifted slightly to lower energies in the supported segments. This suggests that the damping of the excited states involved in the resonant Raman process for the supported carbon nanotube in Fig. 11.18 is greater than the damping for the excited states of the suspended segments of the carbon nanotube, consistent with the observation of localized doping of graphene and carbon nanotubes supported on oxide substrates. These results also suggest that the Raman scattering efficiencies of graphite may underestimate the Raman scattering efficiencies of graphene and carbon nanotubes when they are well isolated from substrate and overlayer effects.

The results of Fig. 11.18 provide a traditional example of resonant Raman scattering with the strong dependence of the Raman scattering intensity on the excitation wavelength. The Raman spectra of graphene based materials also demonstrate a more subtle example of resonance Raman scattering. Both the D and 2D lines of graphene, graphite and carbon nanotubes show dispersion with the Raman energy depending on the excitation wavelength. The Raman energies of the D and 2D lines are determined by a combination of the phonon dispersion, and the electronic dispersion through the satisfaction of a double resonance condition as described earlier and illustrated schematically in Figs. 11.4(c)–(d). The resonance condition means that the intensity of the Raman scattering will depend on the lifetimes of the excited states involved in the double resonance process as suggested in Fig. 11.15.

The results in Figs. 11.13–11.18 show that the careful choice of excitation energies, use of scattering geometries which maximize the fields at the carbon nanotube or graphene layer, and environments which minimize defect scattering of optically excited carriers make the use of Raman scattering to study single graphene layer, and single carbon nanotube devices practical.

11.6. MATERIALS IDENTIFICATION

11.6.1. Determination of the Layer Number in Few-Layer Graphene

The electronic properties of few-layer graphene depend strongly on the exact layer number: Single-layer graphene is a zero-gap semiconductor with conical bands. AB-stacked bilayer graphene is a zero-gap semiconductor with hyperbolic bands, and a bandgap can be opened by applying a perpendicular electric field that breaks the inversion symmetry of the bilayer. Finally, tri-layer and thicker few-layer graphene are semimetals with a small band-overlap between conduction and valence bands. Therefore, it is essential that the layer number of few-layer graphene, exfoliated from graphite, or grown by various methods on substrates, is determined accurately. Raman spectroscopy can contribute as a microscopic materials characterization tool in various ways:

Figure 11.19 shows the G'-band spectra for single-layer, AB-stacked bilayer, and thicker few-layer graphene as well as graphite.[44] The most apparent changes arise when going from single-layer graphene to bilayer graphene. In single-layer graphene the G'-band is a single Lorentzian peak with higher intensity than the G-band. In bilayer graphene, this peak splits into a broadened peak that can be de-convoluted into four components and its overall intensity is reduced. The G'-peak in thicker few-layer graphene and graphite can be fitted with two components. The four components of bilayer graphene have been shown to arise from the altered electronic structure of bilayer graphene as compared to single-layer graphene.

The evolution of the G'/2D structure from a sharp line in graphene which is stronger than the first order G line of graphene, to the broadened feature in graphite which is weaker than the first order G line of graphite is largely completed at the 5−7 layer level. It reflects the importance of double resonance contributions to the Raman cross sections of

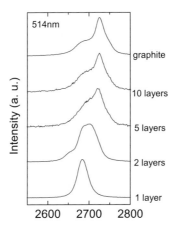

Fig. 11.19   Changes in the 2D-band lineshape with increasing graphene layer number. The single-layer graphene 2D-band can be fitted by one Lorentzian, the bilayer graphene 2D-band consists of four components, and graphite can be fitted with two Lorentzians. Reproduced from Ferrari *et al.*[44]

the second order Raman processes of these materials since in most other materials, the two phonon Raman spectra are broad and resemble the autocorrelation of the phonon density of states. The sharp $G'/2D$ line in graphene reflects both the phonon dispersion curve of graphene and its simple electronic bandstructure with its single layer unit cell. The broadened $G'/2D$ line in bilayer graphene and multi-layer graphene including graphite shows how the more complicated electronic structure of stacked graphene is manifested in the double resonance Raman spectrum.

The above method for determining the layer number works well for exfoliated graphene, where the AB stacking order from natural graphite is preserved. In few-layer graphene grown on silicon-carbide (SiC) however, the stacking order is usually not of the AB type, which leads to largely decoupled electronic layers with strong, single-Lorentzian G' lineshapes. Therefore a different method is needed to measure the layer number of graphene on SiC. The SiC Raman attenuation method[45] (Fig. 11.20) makes use of the fact that graphene absorbs about $\alpha = 2.3\%$ of the light intensity per pass and layer for light in the infrared and visible. The SiC Raman signal from a SiC sample with a graphene overlayer is therefore attenuated because the light has to pass twice through all the graphene layers (The incident light on

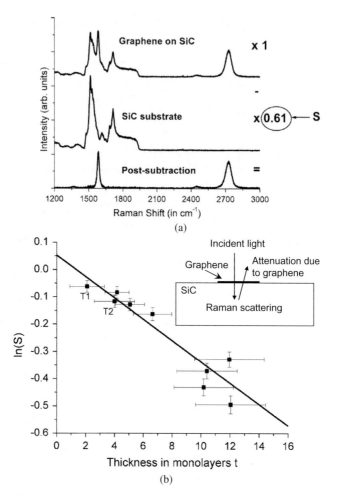

Fig. 11.20 Thickness determination of epitaxial graphene by the substrate Raman attenuation method. (a) Raman spectra of few-layer graphene grown on SiC, Raman spectrum of the bare SiC substrate, and subtracted spectra after scaling the bare substrate spectrum by a factor $S = 0.61$, corresponding to $n \sim 11$ to 12 layers. (b) Plot of $\ln(S)$ versus the sample thickness for different samples. The solid line is a fit to $S = I/I_0 = \exp(-2n\alpha)$ yielding $\alpha = 2.0\%$. Inset: Schematic of the attenuation of the substrate Raman signal. Light has to pass twice through the graphene overlayer. Reproduced from Shivaraman *et al.*[45]

the way into the substrate, and the SiC Raman scattered light on the way back to the microscope objective). The layer number can then be determined from $I/I_0 = \exp(-2n\alpha)$, where I is the SiC Raman intensity of the sample, I_0 is the Raman intensity on a standard SiC sample without the graphene overlayer, n is the graphene layer number, and α is the graphene absorption coefficient. Experimental values of $\alpha = 2.3\%$ and $\alpha = (2.0 \pm 0.2)\%$ have been measured from optical transmission experiments[46] and the Raman study (Fig. 11.20(b)).[45]

11.7. DEVICE CHARACTERIZATION

11.7.1. Temperature Measurements in Biased Devices

Power dissipation is a major concern in current high-performance electronic devices and reliable methods to measure the spatially resolved temperature will be critical to the development of any future technology. In the following we describe two microscopic methods to measure the temperature of a graphene FET by Raman spectroscopy.

(1) The Stokes/anti-Stokes method makes use of the fact that the rate of the anti-Stokes process, where a phonon is destroyed, is proportional to the occupation number of that phonon, n, while the Stokes process, where a phonon is generated, is proportional to $(n + 1)$:[24]

$$R_{\text{Anti-Stokes}} \sim n\left(E_L + \hbar\omega\right)^3 |\chi_{AS}(E_L)|^2$$

$$R_{\text{Stokes}} \sim (n + 1)\left(E_L - \hbar\omega\right)^3 |\chi_S(E_L)|^2$$

Here E_L and $\hbar\omega$ are the laser and phonon energies, and χ_S and χ_{AS} are the Stokes and anti-Stokes Raman susceptibilities. In the Bose-Einstein statistics, the ratio $(n + 1)/n = \exp(\hbar\omega/k_b T)$ is related to temperature T, and therefore the Stokes/anti-Stokes ratio can be used to determine temperature:

$$\frac{R_S}{R_{AS}} = \exp\left(\frac{\hbar\omega}{k_b T}\right)\left(\frac{E_L - \hbar\omega}{E_L + \hbar\omega}\right)^3 \left(\frac{\chi_S(E_L)}{\chi_{AS}(E_L)}\right)^2$$

Time-reversal symmetry requires that $\chi_{AS}(E_L) = \chi_S(E_L + \hbar\omega)$. For graphene, the Raman susceptibility is slowly varying with the laser energy, as long as the laser energy is far enough away from the π Plasmon resonance at around 3.5 eV. Therefore, the factor with the Raman susceptibilities can be neglected and the Stokes/anti-Stokes ratio determined at a single laser energy. In carbon nanotubes, and in graphene at high laser energies, the Raman susceptibility can change rapidly due to electronic resonances. In such cases, the Stokes and anti-Stokes intensities need to be acquired at two different laser energies $E_{L,\text{Anti-Stokes}} = E_{L,\text{Stokes}} - \hbar\omega$ so that the factor with the Raman susceptibilities becomes unity:

$$\frac{R_S(E_L)}{R_{AS}(E_L - \hbar\omega)} = \exp\left(\frac{\hbar\omega}{k_b T}\right)\left(\frac{E_L - \hbar\omega}{E_L}\right)^3.$$

Strictly speaking, the Stokes/anti-Stokes method determines occupation numbers, not temperatures, but in thermal equilibrium, these quantities are equivalent. In thermal non-equilibrium, phonon populations do not obey the Bose-Einstein statistics, and the derived temperatures need to be understood as "effective" temperatures that are related to the occupation number of that particular phonon mode only. In the next section we will show an example of a non-equilibrium phonon population in electrically-biased carbon nanotubes.

(2) As mentioned in Sec. 11.3.2, a rise in temperature is not only associated with changes in phonon occupation numbers that lead to Raman intensity changes, but also with changes in the phonon frequencies themselves. Calizo *et al.*[22] have experimentally determined the graphene G and 2D phonon frequencies as a function of temperature and found a linear relationship for both: $\Delta\omega_G/\Delta T = -0.016\,\text{cm}^{-1}/\text{K}$ and $\Delta\omega_{2D}/\Delta T = -0.034\,\text{cm}^{-1}/\text{K}$. Since the 2D band is a second-order Raman band, the softening is about twice the G-band softening. In addition, unintentional doping affects the G-band frequency more strongly than the 2D band,[18] so that the 2D band is the best choice for these kinds of temperature measurements. In Fig. 11.21 we show the spatially-resolved Raman 2D-band frequency and related temperature in a graphene field-effect transistor at different electrical power levels.[47] Most of the electrical power is dissipated into the Si/SiO$_2$ gate stack, but the images also reveal that the contacts act as heat-sinks. (see Ref. 47 for a detailed analysis of the heat flow).

11.7.2.   Hot Carrier Effects, Non-Equilibrium Phonon Distributions

In carbon nanotubes under high bias, Raman spectroscopy has been used to measure the effective temperatures of four different phonons: RBM,

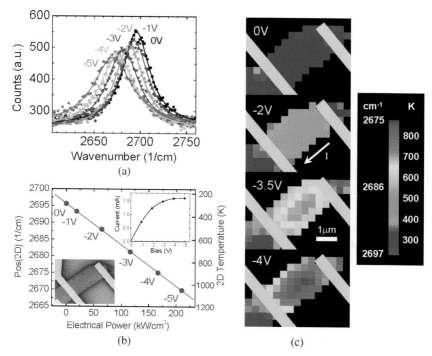

Fig. 11.21   Temperature distribution in a biased graphene field-effect transistor. (a) Raman 2D-band measured in the middle of the graphene device at different bias voltages. (b) 2D-band position as a function of dissipated electrical power. Inset: I-V_d characteristic (top) and SEM image (bottom) of the device. (c) Spatially-resolved images of the 2D-band position (and derived temperature). Reproduced from Freitag *et al.*[47]

G, zone Boundary optical phonon (K), and intermediate-frequency phonon (IFP).[48] Since the RBM and G-band phonons are Raman-active $k = 0$ phonons, their effective temperatures can be measured by the Stokes/anti-Stokes method introduced in Sec. 11.7.1. The K-point optical phonon and the IFP phonons have finite momentum, and thus they are not Raman active. However, their effective temperature can be inferred indirectly through Raman spectroscopy: (1) The primary decay-products of a G-band phonon are two IFP phonons with opposite momentum. Therefore, the lifetime of G-band phonons and width of the G-band depends on the occupation number of the IFPs. Similarly (2) the electronic excitation (an exciton associated with a 1-dimensional carbon nanotube sub-band) decays mainly through a series of K-point phonon emissions. Thus the lifetime and width of

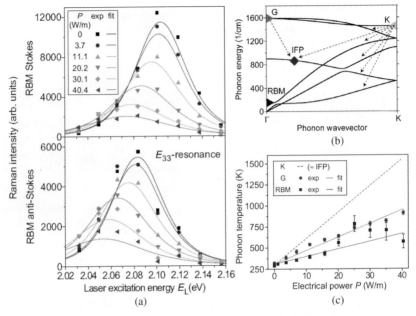

Fig. 11.22   Non-equilibrium phonon distribution in an electrically-biased carbon nanotube. (a) Resonant Raman spectroscopy of the RBM. Stokes and anti-Stokes intensities are shown for different laser energies and electrical power levels. The fitted profiles show the evolution (broadening and shift) of the E_{33} level with increasing temperature of the K-point phonon (which is the main decay product of the exciton). (b) Schematic of the phonon dispersion in a carbon nanotube with the G-band, RBM, K-point, and IFP phonons highlighted. Dominant decay pathways are indicated. (c) Extracted temperatures of the G-band, RBM, K-point, and IFP phonons in the biased carbon nanotube as a function of electrical power. Reproduced from Steiner *et al.*[48]

the exciton is determined by the K-point phonon temperature. Since resonant Raman spectroscopy is sensitive to the electron-phonon (or exciton-phonon) interaction, it can be used to map the exciton resonance (see Sec. 11.5). Figure 11.22(a) shows the resonant Raman profile of the RBM in the biased nanotube at different electrical excitation powers. The profile broadens and shifts with increasing K-point phonon temperature. Similarly, the G-band phonon itself broadens and shifts with increasing IFP temperature. The effective temperatures of all four phonons are shown in Fig. 11.22(c). All phonon temperatures increase roughly linearly with electrical power, however, the K and IFP phonons are significantly hotter than the G and RBM phonons. The discrepancy is explained by predominantly electronic pumping of the K-point phonon due to its large electron-phonon coupling strength and energy significantly below the G-band phonon's, and by a phonon-decay bottleneck at the long-lived IFPs which prevents thermalization with the acoustic phonon bath as represented by the RBM.

11.7.3. Charge Density in Biased Carbon Nanotube Field-Effect Transistors

In Sec. 11.3.1 we have shown that due to strong electron-phonon coupling in graphene, a changing Fermi level can alter the width and energy of the G-band phonons.[14–18] The same holds true for metallic carbon nanotubes, because they too have a linear low-energy electronic dispersion and no bandgap.[19,20] When the Fermi level is shifted more than $|E_F| > \hbar\omega/2$ upon electrostatic gating, the G-band phonon can no longer decay into electron-hole pairs due to the Pauli Exclusion Principle, which increases the phonon lifetime and decreases the width of the Raman line. In addition, the phonon energy is affected due to changes in phonon renormalization upon doping.

In semiconducting carbon nanotube field-effect transistors, the bandgap prevents the coupling of phonons with low-energy excitations of electrons. Therefore the widths of the Raman G-lines in un-doped semiconducting carbon nanotubes are narrower than in their metallic counterparts and they stay essentially constant upon doping (Fig. 11.23(b)). However, as the Fermi level moves into conduction or valence bands, the renormalization of the phonon energy due to emission and absorption of virtual electron-hole pairs is affected.[19] (Virtual excitations may have energies higher than $\hbar\omega$). This can be seen in Fig. 11.23(a) and (c), where the G-band of the semiconducting nanotube moves to higher energies upon electrostatic doping with holes. Figure 11.23(e) shows the Raman G-band of another semiconducting carbon nanotube field-effect transistor that was biased not only by a gate voltage (3 V) but also by a drain voltage (6 V), and a current of $8\,\mu$A was flowing as a result.[49] It is apparent that the Raman G-band is stiffened in the center of the device as compared to the area next to the contact. Carbon nanotubes are Schottky-barrier field-effect transistors, and the non-uniform doping, shown in the schematic in Fig. 11.23(f), can explain the shifting G-band position. These shifts in the phonon energy (and changes of linewidth

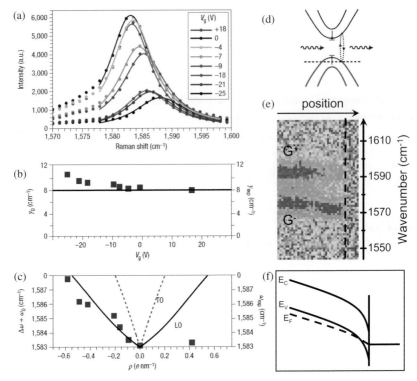

Fig. 11.23 Doping and phonon renormalization in semiconducting carbon nanotubes. (a) G-band Raman spectrum of a semiconducting carbon nanotube field-effect transistor under varying gate voltages. (b) G-band width as a function of gate voltage. (c) G-band position as a function of doping. (d) Schematic of the phonon renormalization for the case of the Fermi level in the valence band. (e) Spatial map of the Raman G-band of another semiconducting carbon nanotube under bias in the vicinity of a contact (G^+ and G^- are visible). The dashed line indicates the position of one of the contacts. (f) Schematic of the band-bending and Fermi level in the biased carbon nanotube field-effect transistor. (a–d) reproduced with permission from J. C. Tsang et al.[19] (e–f) reproduced from M. Freitag et al.[49]

in metallic carbon structures) as a function of the carrier density can be useful in the characterization of these materials in FET structures under applied gate fields.

11.8. CONCLUSION

We have shown that Raman spectroscopy can provide valuable information about carbon nanotubes and graphene, and field effect transistors where these materials are used as the channels. Although the amount of active material in these samples can be small compared to other systems where Raman scattering has been previously used as a characterization tool, the Raman scattering efficiencies are sufficiently large for the Raman signals to show reasonable signal to noise ratios. This is especially so for sample geometries similar to those pioneered by Geim *et al.* where the electromagnetic response of the carbon layer can be enhanced by multilayer interference effects. Raman scattering can be used to determine important material parameters such as the diameter of a carbon nanotube, the number of layers in a graphene sample, the doping profile along a channel, temperature, strain, and disorder. These measurements can be made even when the carbon nanotube or graphene structure is covered by a transparent overlayer as is often the case in a packaged working device.

We have discussed several examples of problems that can cause anomalies in the electrical performance of devices using carbon nanotubes or graphene and which can be resolved by Raman scattering. Problems in the electrical behavior of a carbon nanotube FET can arise from a failure to properly identify whether the nanotube used in the device is a semiconductor or a metal. This can be resolved by measurements of its diameter by the energy of the RBM and the resonances of the Raman spectra both of which can be fit to theoretical determinations of the diameter and critical points of the electronic bandstructure to unambiguously determine the tube chirality. The electrical behavior of a graphene FET depends critically on the number of layers in a graphene film and this can be measured by the energies and relative intensities of the G and G′/2D lines. Devices can show anomalous gate voltage characteristics due to accidental doping of the channel during fabrication, which changes both the positions of the G and G′/2D lines and their relative intensities. The processing of the carbon nanotube or graphene can reduce the mobility of the channel material because it introduces disorder which can be detected through the intensity of the D line. Even suspended devices which show the highest mobilities have regions resting on substrates and under overlayers whose electronic properties can be altered by local strains which can be measured by shifts in the phonon frequencies, intensities and splittings. Anomalies in high field behavior of the carriers in the channels due to the generation of hot phonon populations under finite applied voltages can be quantified through their signatures in the Stokes and anti-Stokes Raman intensities of both thermal and non-thermal distributions of the $k = 0$ and $k = K$ point phonons and also their energies.

In this discussion we showed what can be learned from conventional Raman measurements at one or two excitation wavelengths. Raman spectra obtained in resonant Raman measurements also determine the energies of the van Hove singularities in carbon nanotubes. The independent determination of the tube diameter from the energy of the RBM line and the positions of the van Hove singularities of the electronic structure of the carbon nanotube provides the most accurate determination of the chirality. The spectral widths of the van Hove singularities provide direct data on the relaxation of the excited states and the impact of the local environment on the electronic properties of the nanotube. While the resonant Raman spectra of graphene in the visible shows no sharp structure, the dependence of the resonant Raman intensities on the excited state lifetimes means that the Raman intensities of the G and G'/2D lines contain information about the damping of the electronic excited states which can be important in obtaining the highest possible mobilities from these devices.

While all of the Raman studies discussed here have involved static measurements, time resolved Raman spectroscopy has been widely used to understand the dynamical properties of semiconductors and metals. Time resolved Raman scattering have studied the dynamics of the optical phonons of graphite and carbon nanotubes in the temporal domain and directly measure the relaxation of these excitations without the uncertainties associated with the conversion of linewidth into lifetime.[50] The similarities in the intensities of the Raman scattering of interference enhanced Raman scattering of graphene and of bulk graphite suggest that similar measurements can be made for graphene, providing the opportunity to study how substrates and overlayers change the phonon relaxation processes, and how this will depend on different substrates and overlayer treatments. Such results will be important in qualifying new graphene and carbon nanotube structures for further device applications. While the interference enhancement of the Raman signals can change the temporal response of the optical system, the low reflectivity of the graphene and the Si substrates commonly used in these experiments should produce relatively low Q cavities, and the slowing of the temporal response due to cavity effects should be weak. It should be mentioned that pump-probe studies of hot carrier dynamics of graphene using short pulsed lasers have recently been reported.[51]

Finally, while the focus of this article has been on Raman spectroscopy, it should be remembered that Raman spectroscopy is one element of a family of phenomena of secondary light emission. These effects can be easily studied by experimental equipment used for static and time dependent Raman scattering measurements. This includes hot carrier emission processes involving broad band emission extending beyond one and two phonon energies from the excitation energy.[52] These processes have been used to study the dynamics of hot carriers in semiconductors such as Si and GaAs. Multi-order Raman scattering has been reported from individual carbon nanotubes[53] for energies up to 1 eV below the excitation energy and involving as many as six optical phonons. The study of such processes in carbon nanotubes and graphene should provide very detailed information on carrier relaxation processes in these materials and their unusual electronic bandstructures. Such studies can involve both the static excitation of carriers by either optical or electrical means as well as their pulsed excitation.

We have mentioned that Raman spectroscopy is capable of submicron spatial resolution. In fact, the combination of excitation wavelengths of the order of 400 nm and the use of index matched optical objectives and protective packaging layers with index of refractions in the range of 1.5 suggests that spatial resolution as small as 200 nm can be achieved currently. If, as is usually the case, the protective packaging layer is optically transparent, then Raman spectroscopy can be applied to a working device without any extra processing. This is in contrast to e-beam techniques for circuit analysis which require that the voltage carrying layer of the device be directly exposed to the electron beam. For working packaged devices, this requires the removal of the packaging layer and other layers such as those associated with wiring layers before the e-beam probe can be applied. Since this destructive sample preparation can change the properties of the device, ideally it should be minimized. Raman spectroscopy with its sub-micron spatial resolution in buried devices with minimal additional processing and its sensitivity to a variety of physical parameters of carbon based IC technologies is a useful tool for the characterization of carbon nanotube and graphene devices.

REFERENCES

1. International Technology Roadmap for Semiconductors 2007 Edition. Available: http://www.itrs.net/Links/2007ITRS/Home2007.htm
2. L. Pauling, *The Nature of the Chemical Bond*. 1960: Cornell University Press, Ithaca, NY.
3. sematech.org/docubase/document/4532atr.pdf
4. C. Thomsen, S. Reich, *Raman Scattering in Carbon Nanotubes* in *Light Scattering in Solid IX*, M. Cardona, R. Merlin, Editors, Springer: Berlin/Heidelberg, pp. 115–234 (2007).
5. M. S. Dresselhaus, *et al.*, *Nano Lett.* **10**, 751 (2010).
6. M. Cardona, *Resonance Phenomina*, in *Light Scattering in Solids II*, M. Cardona, G. Guntherodt, Editors, Springer: Berlin, p. 19 (1982).
7. R. Loudon, *Proc. R. Soc. Lond. A* **275**, 218 (1963).
8. J. Menendez, G. S. Spenser, G. H. Loechelt, *Proceedings of the Symposium on Diagnostic Techniques for Semiconductor Materials and Devices*, D. K. Schroeder, J. L. Benton, P. RaiChoudhary, Editors, Electrochemical Society Proc. pp. 217–227 (1994).
9. D. M. Basko, *New. J. Phys.* **11**, 095011 (2009).
10. M. S. Dresselhaus, P. Avouris, *Introduction to Carbon Materials Research*, in *Carbon Nanotubes*, M. S. Dresselhaus, Ph. Avouris, Editors, Springer, Berlin, pp. 1–9 (2001).
11. M. Mohr, *et al.*, *Phys. Rev. B* **76**, 035439 (2007).
12. C. Thomsen, S. Reich, *Phys. Rev. Lett.* **85**, 5214 (2000).
13. I. DeWolfe, *Semicond. Sci. and Tech.* **11**, 139 (1996).
14. S. Piscanec, *et al.*, *Phys. Rev. Lett.* **93**, 185503 (2004).
15. S. Piscanec, *et al.*, *Phys. Rev. B* **75**, 035427 (2007).
16. J. Yan, *et al.*, *Phys. Rev. Lett.* **98**, 166802 (2007).
17. S. Pisana, *et al.*, *Nature Mater.* **6**, 198 (2007).
18. A. Das, *et al.*, Monitoring dopants by Raman scattering in an electrochemically top-gated graphene transistor. *Nature Nanotech.* **3**, 210 (2008).
19. J. C. Tsang, *et al.*, *Nature Nanotech.* **2**, 725 (2007).
20. H. Farhat, *et al.*, *Phys. Rev. Lett.* **99**, 145506 (2007).
21. N. Bonini, *et al.*, *Phys. Rev. Lett.* **99**, 176802 (2007).
22. I. Calizo, *et al.*, *Appl. Phys. Lett.* **91**, 071913 (2007).
23. Y. Zhang, *et al.*, *J. Phys. Chem. C* **111**, 14031 (2007).
24. A. Compaan, H. J. Trodahl, *Phys. Rev. B* **29**, 793 (1984).

25. T. M. G. Mohiuddin, *et al.*, *Phys. Rev. B* **79**, 205433 (2009).
26. M. Huang, *et al.*, Phonon softening and crystallographic orientation of strained graphene studied by Raman spectroscopy. *Proceedings of the National Academy of Sciences* **106**, 7304 (2009).
27. B. S. Elman, *et al.*, *Phys. Rev. B* **24**, 1027 (1981).
28. F. Tuinstra, J. L. Koenig, *J. Phys. Chem.* **53**, 1126 (1970).
29. L. G. Cancado, *et al.*, *Appl. Phys. Lett.* **88**, 163106 (2006).
30. J. A. Robinson, *et al.*, *Nano Lett.* **9**, 2873 (2009).
31. K. Sinha, J. Menéndez, *Phys. Rev. B* **41**, 10845 (1990).
32. R. J. Nemanich, C. C. Tsai, G. A. N. Connell, *Phys. Rev. Lett.* **44**, 273 (1980).
33. K. S. Novoselov, *et al.*, *Science* **306**, 666 (2004).
34. C. Casiraghi, *et al.*, *Nano Lett.* **7**, 2711 (2007).
35. Y. Y. Wang, *et al.*, *Appl. Phys. Lett.* **92**, 043121 (2008).
36. D. Yoon, *et al.*, *Phys. Rev. B* **80**, 125422 (2009).
37. Z. H. Ni, *et al.*, *ACS Nano* **3**, 569 (2009).
38. K. I. Bolotin, *et al.*, *Solid State Communications* **146**, 351 (2008).
39. X. Du, *et al.*, *Nature Nanotech.* **3**, 491 (2008).
40. C. Fantini, *et al.*, *Phys. Rev. Lett.* **93**, 147406 (2004).
41. A. Jorio, *et al.*, *Phys. Rev. Lett.* **86**, 1118 (2001).
42. S. M. Bachilo, *et al.*, *Science* **298**, 2361 (2002).
43. M. Steiner, *et al.*, *Appl. Phys. A* **96**, 271 (2009).
44. A. C. Ferrari, *et al.*, *Phys. Rev. Lett.* **97**, 187401 (2006).
45. S. Shivaraman, *et al.*, *J. Electron. Mat.* **38**, 725 (2009).
46. R. R. Nair, *et al.*, *Science* **320**, 1308 (2008).
47. M. Freitag, *et al.*, *Nano Lett.* **9**, 1883 (2009).
48. M. Steiner, *et al.*, *Nature Nanotech.* **4**, 320 (2009).
49. M. Freitag, J. C. Tsang, P. Avouris, *Phys. Stat. Sol. B* **245**, 2216 (2008).
50. H. Yan, *et al.*, *Phys. Rev. B* **80**, 121403 (2009).
51. R. W. Newson, *et al.*, *Opt. Express* **17**, 2326 (2009).
52. J. A. Kash, J. C. Tsang, *Light Scattering and Other Secondary Emission Studies of Dynamic Processes in Semiconductors*, in *Light Scattering in Solids VI*, M. Cardona, G. Guntherodt, Editors, Springer Verlag: Berlin, pp. 423–515 (1991).
53. F. Wang, *et al.*, *Phys. Rev. Lett.* **98**, 047402 (2007).
54. F. Cerdeira, *et al.*, *Phys. Rev. Lett.* **57**, 3209 (1986).
55. D. M. Basko, *Phys. Rev. B* **78**, 125418 (2008).

12

SINGLE NANOWIRE PHOTOELECTRON SPECTROSCOPY

Carlos Aguilar*
MIT Lincoln Laboratory
244 Wood St., Lexington, MA 02420, USA

Richard Haight
IBM T. J. Watson Research Center
Yorktown Heights, NY 10598, USA

*This work is sponsored by the Air Force under Air Force Contract #FA8721-05-C-0002. Opinions, interpretations, recommendations and conclusions are those of the authors and are not necessarily endorsed by the United States Government.

12.1.   INTRODUCTION

Photoelectron spectroscopy has become a powerful and indispensable tool for studying the electronic structure and properties of materials.[1] When energetic photons in the vacuum ultraviolet region of the spectrum (6–150 eV) are absorbed by a material, the liberated electrons carry information on the surface and bulk electron energies as a function of momentum thereby allowing one to map the electronic band structure of the material. In addition to band structure mapping, the work function of the material and the location of the Fermi level within the band gaps of semiconductors, and even thin insulator layers can be determined. Using photons in the energetic region beyond vacuum ultraviolet (soft and hard X-rays, 150–1000 eV) the chemical state of the surface and bulk atoms in both elemental and compound solids can be determined by measuring the shifts of the atomic core levels.

Beyond these fundamental measurements, which can be carried out with resonance lamps, X-ray tubes or at synchrotron facilities, the advent of photoemission with ultrafast lasers has opened up new vistas in the study of material electronic properties in the excited state.[2] Dynamic electronic behavior such as electron-phonon scattering,[3] surface recombination,[4] band gap renormalization,[5] the formation of hydrogenic-like image states,[6] exciton creation[7] and photovoltage shifts[8] have all been studied with femtosecond laser photoemission spectroscopy.

Given the extraordinary analytical power of photoelectron spectroscopy, it is worth considering the challenge of applying these techniques to the study of nanostructures both as ensembles, but perhaps more importantly, on an individual basis. The study of ensembles of nanostructures provides broad information but the typical distributions of size, shape and composition often wash out detials of critical electronic properties.

Typically, achieving higher spatial resolution with optical spectroscopies (of which photoelectron spectroscopy is a subset) for the study of nanostructures involves improving the focal properties of the excitation light. For high energy photons some examples of focusing optics include curved metal reflective optics,[9] bent crystals,[10] Fresnel Zone plates,[11] glass capillaries[12] and X-ray multilayer mirrors.[13] These approaches have reached the 10's to 100's of nanometer focal diameters and will certainly continue to shrink with improvements in optical components, although fundamental diffraction ultimately limits the achievable resolution. Alternatively, in the case of photoelectron spectroscopy, sophisticated electron lenses can be used to achieve high spatial resolution as in the case, for example, of PEEM[14] (photoelectron emission microscopy) which has achieved resolutions in the 10–30 nm range (see Chap. 5). LEEM (low energy electron microscopy) of which PEEM is a variant, has achieved close to 2 nm spatial resolution at the surfaces of materials and is described in detail in Chap. 4 of this handbook.

A different, and arguably simpler, approach to achieving high spatial resolution involves isolating the nanostructure under consideration and using modest focusing optics to study its electronic structure. This approach has been successfully used in a series of beautiful Rayleigh scattering experiments[15] to study carbon nanotubes (CNT) with diameters on the order of 1 nm. In this experiment, CNTs are grown or deposited across a gap region etched into a suitable substrate such as Si. If the gap is sufficiently large (≥ 10 microns) and the distances between each individual CNT is similarly large, it is possible then to focus a light beam or pulse onto the CNT and monitor the scattered light. While these experiments were carried out on CNTs, the approach can be similarly extended to other nanowires (NW) and with other spectroscopies, such as photoelectron spectroscopy discussed here.

This chapter is organized as follows. We will discuss the fundamental aspects of photoelectron spectroscopy utilized in the study of nanostructures, and in particular semiconductor NWs using low photon energies. Lower photon energies in particular are useful since it is possible to optimize the absorption length[16] of the incident light and the electron inelastic mean-free-path[17] or escape depth to match the dimensions of the NW. We will then discuss carrying out photoemission using femtosecond lasers and the advantages of this approach. Focusing energetic light onto nanostructures requires geometries that allow for introduction of the light and escape of the emitted electrons; we will therefore discuss the unique forward geometry used in our NW photoemission experiments. We will describe, in some detail the apparatus, patterned substrate preparation, NW growth and chemical functionalization of the semiconductor NWs studied.

Following the introductory discussion we will describe experimental examples of single NW photoelectron spectroscopy. We begin with studies of individual Si and Ge NWs[18] in Sec. 12.3 that reveal their electronic properties and the differences from wire to wire based upon their shape and diameter. In addition we show that fundamental optical properties in the deep UV can be extracted from these measurements. We then go on to discuss doping of individual NWs, in Sec. 12.4 and describe the method by which photoemission can be used to determine NW doping levels through the accurate location of the Fermi level position within the NW band gap. A further example is given in Sec. 12.5, where Si NWs grown with Al catalysts, as compared with the more traditional Au, reveal significant Al incorporation into the NW matrix, but exhibits only partial activation that increases with post-growth anneals.[19] Chemical functionalization of individual Si NWs[20] as a means of controllably modifying the Fermi level position is described in Sec. 12.6. In Sec. 12.7, we describe measurements of single walled carbon nanotubes (CNT), which represent the smallest individual solid structures measured with photoelectron spectroscopy. Finally in Sec. 12.8 we discuss measurements on ZnO NWs,[21] both pristine and chemically functionalized and describe the role of that both oxygen vacancies and surface chemical treatments play in doping of the wires.

12.2. SINGLE NW PHOTOEMISSION FUNDAMENTALS

The fundamental understanding of electron excitation in a solid by an energetic photon can be traced to Einstein[22] who was awarded Nobel prize in 1921 for delineating this simple but exceedingly important process. In

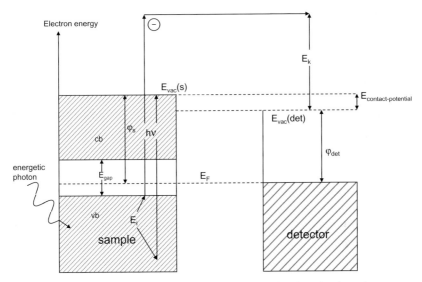

Fig. 12.1 Schematic showing fundamental energetic considerations in photoelectron spectroscopy. Connection of the sample to the detector equilibrates Fermi levels, but can introduce a contact potential.

general, the photoemission process can be resolved into a sequence of events which begin with the absorption of an energetic photon that promotes the electron to a state above the vacuum level of the solid (Fig. 12.1). The vacuum level is that energy to which a liberated electron exists outside the solid with zero kinetic energy. Therefore the kinetic energy of the electron, measured by an electron analyzer is given by:

$$E_{kin} = h\nu - (E_{vac} - E_i) + E_{cp} \tag{12.1}$$

where E_i is the intial state energy the electron occupies within the solid, typically in the valence band of the solid, but also within the conduction band if the electron exists there due to photoexcitation or doping. If the electron is at the Fermi level, as in the case of a metal, $(E_{vac} - E_i)$ would correspond to the work function, φ, of the solid. E_i is the binding energy with respect to the Fermi level and E_{cp} is the contact potential created when the analyzer is electrically connected to the sample.

In photoelectron spectroscopy, the Fermi level is the fundamental reference level. A useful paper on these considerations by Cahen and Kahn[23] provides a thorough discussion of photoemission energetics but we capture the salient points here. When two unrelated solids are brought into intimate electrical contact (implying that electrons can flow from one solid to the other) electrons will flow from the solid with the smaller work function to the solid with the larger work function until the Fermi levels of the two solids equilibrate or are the same. The result of this charge transfer is that an electrostatic potential is then set up between the two materials, called a contact potential. Interestingly this occurs between two materials in intimate contact or between a solid under study and the electron energy analyzer used to study it with photoemission. In the case of the material couple, the field is internal to the coupled solid while for a solid and electron detector the field is between the solid and the detector. This field can accelerate or decelerate the photoemitted electrons and must be taken into account during the experiment. Nonetheless, the Fermi level is then the key reference level and all energetic changes experienced by the sample during the experiment are referenced to this level. This level is easily established to high precision by evaporating a metal film in electrical contact with the sample and then generating a photoemission spectrum to reveal the sharp Fermi edge associated with the cutoff of electronic occupancy dictated by Fermi statistics.[24]

12.2.1. Electron Escape Depth

Another set of important considerations, particularly relevant to photoemission from a NW or other nanoscale structure involves the photoemission photon energy and the inelastic mean-free path of the emitted electrons.[17] Since the NWs we are interested in studying are in the range of tens to hundreds of nanometers in diameter, photoelectron spectroscopy with low energy photons is particularly advantageous. The absorption of light in a material is governed by the Beer–Lambert law of attenuation given by:

$$I = I_o e^{-\alpha z} \qquad (12.2)$$

where α is the optical absorption coefficient and z is the distance into the solid. A useful tabulation of absorption coefficients for photon energies up to 6 eV, for a range of elemental and compound semiconductors can be found in a paper by Aspnes and Studna.[16] As we will discuss later the absorption of 6 eV photons in Si results in an intensity drop to its e^{-2} value

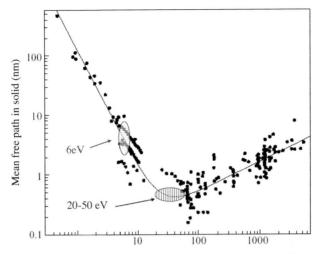

Fig. 12.2 Electron mean free path as a function of kinetic energy of the electron in the solid. For 6 eV electrons the mean free path is approximately 10 nm (from Ref. 17).

in ~10 nm from the surface. The inelastic mean-free path for electrons with an energy of 6 eV above the Fermi level is ~6–10 nm (Fig. 12.2.)

The combination of light absorption and electron mean-free-path, or escape length then provides an interesting length scale; both the surface and bulk electronic structure of a NW whose diameter is ~10's of nanometers can be studied. For high energy photons the situation is a bit more complicated. Samson[9] provides useful data on the absorption of vacuum ultraviolet light in a wide range of materials; the magnitude is highly wavelength dependent due to transitions from atomic core levels to states above the Fermi level. As an example, the transmittance of a 200 nm film of Si climbs from 10% at a photon energy of 25 eV to nearly 70% at 95 eV but plummets above this energy due to strong absorption by the Si 2p core level. Conversely, the mean free path of photoelectrons at these higher energies is relatively short, typically <1 nm. Hence such energetic light is suitable to penetrate a NW with a 10 or 20 nm diameter, but the spectral information will be considerably more surface derived than with 6 eV photons. Therefore we see that it is possible to study both bulk and surface electronic structure of nanoscale objects, and in particular, NWs with a wide range of photon energies.

These considerations are also particularly relevant to photoemission from a single NW because, as we will discuss, geometrical considerations

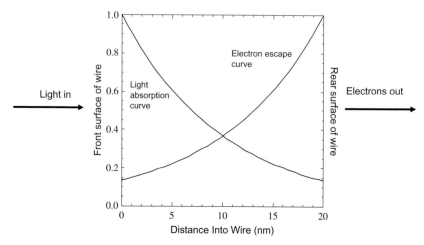

Fig. 12.3   Schematic showing a NW of nominally rectangular cross-section. Photoemission light impinges on the wire from the left (front surface of NW), the intensity experiences an exponential decay with depth into the NW. Photoelectrons emerge from the NW on the right (rear surface of NW); electrons at the rear surface have the highest probability of emerging without being inelastically scattered.

force us to focus light onto one side of the NW with the ensuing photoemitted electrons exiting from the opposite side of the wire (Fig. 12.3). Single NW photoelectron spectroscopy is configured in a similar manner to that utilized in Rayleigh spectroscopy experiments on carbon nanotubes.[15] Figure 12.4 shows schematically how photoemission is carried out on an isolated single NW. A number of considerations contributed to the particular geometry used.

12.2.2.   Nanowire Deposition

To start, suspension of individual NWs across gaps in a conducting substrate is required, allowing the introduction of light and emission of electrons. The Si substrate, shown schematically in Fig. 12.5, was lithographically patterned and etched completely through its thickness to achieve a tapered edge to the slot.[18] Arrays of slots were etched to provide high probability of locating a NW. Fiducials in the form of letters, numbers and markings along the slot edges as well as row/column markers identifying individual slots were used so following NW growth, they could be found in a scanning electron microscope (see Chap 1).

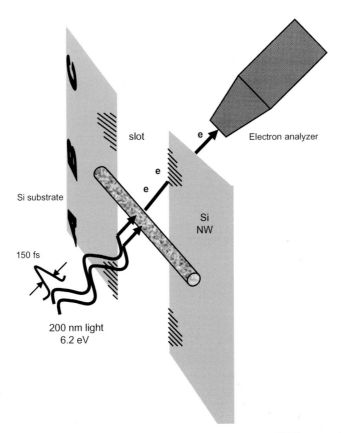

Fig. 12.4   Schematic diagram of single NW photoemission. A single NW is suspended across a gap in the Si substrate. The substrate is dressed with fiducials for identifying the location of the NW. Energetic light is incident on the front of the wire and electrons emerge from the back surface which faces the electron energy analyzer.

Once the substrate with NWs was transferred into the analysis chamber, an optical system integral with the photoemission system was used to locate the NW.

To achieve isolated photoexcitation of the NWs, a high numerical aperture (NA), vacuum compatible microscope objective was used; the high NA achieved a small focal spot diameter but required positioning of the microscope objective within close proximity of the sample. Because of the use of a high NA objective, geometrical considerations dictated the location of the electron detector. While front surface location of the electron analyzer is, at least in principle, possible, only electrons emitted

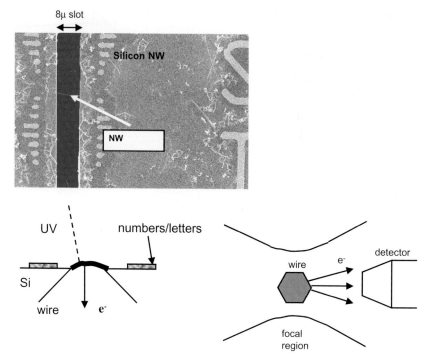

Fig. 12.5 *Upper*: Scanning electron micrograph of a single NW suspended across an $8\,\mu$ slot in Si. *Lower left*: schematic showing side view of a NW spanning a slot. *Lower right*: cross sectional view of focal region showing a faceted NW and emergent electrons.

at large angles relative to the NW surface normal could be studied. An alternative geometry, as shown in Figs. 12.4 and 12.5 locates the analyzer behind the NW. As discussed above, light transmission and electron escape depths, particularly at low photon energies work exceptionally well when studying nanoscale structures such as NWs. In addition, as we will describe later, polarization of the light parallel and perpendicular to the NW is easily managed in this geometry and the large angles of incidence associated with the high NA of the objective further provides control over polarization vectors.

The microscope objective chosen was of the Schwarzchild design (Fig. 12.6). This type of reflective objective was chosen for a number of reasons. Reflective designs such as this avoid chromatic aberrations

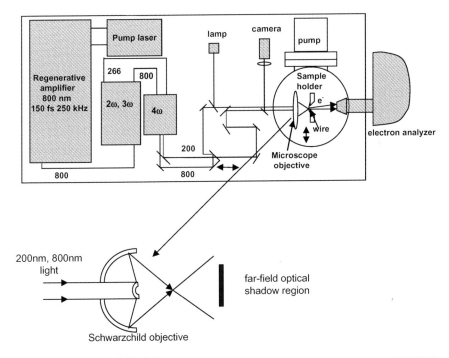

Fig. 12.6 *Upper*: Schematic layout of the femtosecond laser system, harmonic converter, light path and photoemission apparatus. *Middle*: expanded view of the Schwarzchild objective. Beyond the focus, the analyzer is shadowed from any light exposure due to the far field image of the secondary — this is critical to eliminate large background signal from electrons emitted from within the analyzer in this in-line geometry. *Bottom*: Views of the Schwarzchild objective and X–Y piezo stage during construction (*left*) and a side-on view of the analysis chamber after assembly.

which are substantial in refractive systems over the wavelengths used here. This is important when wavelength tunability, pump and probe modalities and imaging issues are considered. As we will describe later in the chapter, optical pumping of a functionalized Si NW at 800 nm was used to evaluate the presence of electric fields (photovoltage shifts) in the wire and the reflective objective provided focusing and good overlap of the 800 nm pump light at the same spot as the 200 nm photoemission light. In addition, 248 nm light integral with the 200 nm and 800 nm beam paths was used for imaging. An interesting benefit of the Schwarzchild design was obtained in the far field; the secondary mirror in the objective created a circular shadow at the entrance of the electron analyzer preventing any light from entering into the lens system of the analyzer. Such light, if it entered the analyzer would generate a substantial background of electrons photoemitted from its interior surfaces.

As described above, a wide range of light sources could be employed for carrying out photoemission on individual NWs, including synchrotrons, resonance lamps and lasers. In the case of synchrotrons, Fresnel zone plates or reflective focusing optics with surfaces coated with multilayers reflective in the vacuum or extreme UV can be used, particularly if cost is not an issue. As mentioned earlier, table-top femtosecond lasers are an attractive source of energetic photons. A wide range of wavelengths can be generated with short pulsed laser sources.

12.2.3. Femtosecond Laser System and Electron Detection

An example of a high repetition rate femtosecond laser source is shown in Fig. 12.6. In this particular apparatus a regeneratively amplified Ti:sapphire laser produces 150 femtosecond pulses of 800 nm light at a high repetition rate, in this case 250 kHz. These 800 nm pulses of light (1.55 eV/photon) must be converted to light sufficiently energetic to photoemit electrons; this is accomplished by first doubling the frequency of the 800 nm to 400 nm (3.1 eV) in an appropriate non-linear crystal such as BBO (beta-barium borate) followed by summing the frequencies of the 800 nm and 400 nm photons in a second crystal to generate 266 nm photons (4.67 eV). Finally residual 800 nm light is summed with 266 nm photons to produce the fourth harmonic at 200 nm (6.2 eV). This light is sufficient to photoemit electrons from the near-valence region of most materials. Higher photon energies can be created through tuning of the source laser to shorter wavelengths or high harmonic approaches. Large quantities of energetic light can be produced through a number of sophisticated approaches such as intracavity high harmonic generation.[25] Other important high repetition rate sources can be utilized such as synchrotrons where tunable photons at 10's of MHz are available. Returning to Fig. 12.6, the 200 nm, 6.2 eV light is then directed into a chamber housing the reflective objective, sample holder and electron analyzer. The sample is manipulated via a 2 axis piezoelectric inchworm stepper in both x and y, while the objective is focused in z with a similar inchworm motor.

Count rates can be estimated in the following analysis. If the pulse of light is focused into a 1 micrometer spot on a 20 nm diameter NW, the rectangular cross-sectional area is then 2×10^{-14} m^2. Typical 6.2 eV pulse energies focused into this area are $\sim 1 \times 10^{-11}$ J/pulse at a repetition rate of 250 kHz. For 6.2 eV photons, this corresponds to 10^7 photons striking this area on the NW per pulse or 2.5×10^{12} photons/sec. For a quantum efficiency of 10^{-3} (electron out/photon in), 2.5×10^9 photoelectrons/sec are emitted into 2π steradians. Since the hemispherical detector used here subtends a solid angle of $\sim 10^{-4}$ steradians we would expect an upper limit of $\sim 10^5$ counts/sec. Typical count rates that we observe experimentally are 10^4–10^5 counts/sec but, as expected, drop with NW diameter so that count rates from a 1.7 nm single wall carbon nanotube are ~ 1000 counts/sec.

These count rates can be substantially increased by using a high power femtosecond laser oscillator to generate the 6.2 eV light since the typical repetition rate of an oscillator is 80 MHz. There are several particular advantages to carrying out photoemission at such high rates. The first is that substantially higher count rates can be achieved with smaller contributions from space charge. Space charge is a serious concern when carrying out photoemission with pulsed sources since each laser pulse emits electron bunches. This is particularly true for photoemission from nanoscale objects such as a NW where the electron bunches are spatially concentrated. Implicit in the count rate discussion above is the necessity for pulse intensity control to avoid distortion of the spectra due to space charge. Pump-probe experiments with higher repetition rate lasers are also more attractive owing to the better signal averaging and lower peak intensities required.

12.3. Si AND Ge NWs

12.3.1. *Photoelectron Spectroscopy*

Semiconductor NWs are of particular technological importance owing to the substantial existing infrastructure devoted to Si electronics, but there is also strong fundamental interest associated with their low-dimensional character. Since control of the NW size can be achieved by a number of growth mechanisms including the widely used vapor-liquid-solid (VLS) process,[26] the potential to study individual wires and the corresponding properties of electrons confined in two dimensions is particular alluring.

Individual NWs with diameters ranging from 20 nm to 400 nm have been isolated and photoelectron spectra near the top of the valence bands have been recorded. With this approach we have been able to extract important fundamental properties of the NWs including the location of the Fermi level within the NW band gap and the work function of H-terminated surfaces. We have also monitored the photoelectron emission intensity as a function of the polarization of the light incident on the NWs and compared this with Mie theory.

Numerous Si and Ge NWs of various diameters, resulting from changes in the VLS growth conditions or from intentional oxidative thinning,[27] were identified and measured with scanning electron microscopy (SEM). Representative micrographs are shown in Fig. 12.5 for Si NWs and Fig. 12.7 for Ge. In the case of Ge NWs, tapered NW needles were observed that extended partly into the slot, although not fully across. Active thinning of Si NWs was achieved by furnace annealing of the sample in an oxygen ambient which formed an SiO_2 shell that was subsequently removed in a hydrofluoric (HF) acid vapor. All samples were exposed to HF vapor to remove residual SiO_2 and to hydrogen terminate the surfaces of the NWs before measurement.

Photoelectron spectra of both Si and Ge NWs are shown in Fig. 12.8. The Si NW possessed a diameter of 20 nm and was collected with light

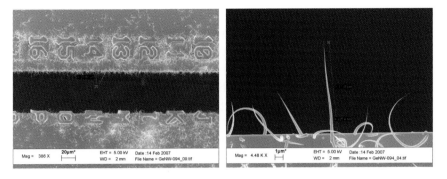

Fig. 12.7   Scanning electron micrographs of tapered Ge NWs extending into the Si slot.

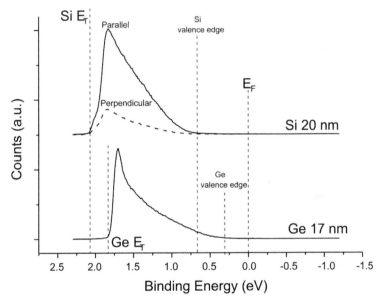

Fig. 12.8   *Top*: Photoelectron spectra showing the emitted electron intensity as a function of binding energy for a 20 nm Si NW excited with 6.2 eV photons polarized parallel (solid) and perpendicular (dashed) to the long axis of the NW. *Bottom*: Photoelectron spectrum of a 17 nm Ge NW. Arrows identify the Fermi level E_F, valence band edge and electron emission threshold, E_T (Ref. 18).

polarized parallel (solid curve) to the wire axis and separately perpendicular (dashed) to the wire axis. In this figure, the Fermi level (E_F) is located at 0 eV binding energy while the NW valence edge is at 0.65 eV. Hence the Fermi level is near the middle of the Si NW band gap, consistent with the undoped nature of the NWs. This location for E_F was observed for all of the Si NWs we studied independent of diameter. We note that for NWs with diameters >10 nm no significant quantum confinement induced band gap changes are expected.[28]

For the Si NWs we studied, the electron emission threshold, E_T, which also determines the work function Φ via the equation $\Phi = 6.2 - E_T$, varied within an energetic span of ± 0.2 eV centered about 2.1 eV. For the particular Si NW in Fig. 12.8, the threshold was located at 2.07 eV and the corresponding value for the work function, $\Phi = 4.13$ eV. A spectrum collected from a Ge wire, near its tip where the diameter was measured to be 17 nm is shown at the bottom of Fig. 12.8. The Ge NW valence band edge is found at 0.31 eV and the threshold energy, $E_T = 1.77$ eV. Since the Ge NWs were grown undoped we again expect, and find, the Fermi level to be near the middle of the Ge band gap. In addition $\Phi = 4.43$ eV for this NW, which is 300 meV larger than for the Si NW discussed above.

The variation of work functions we observed in the Si NWs derives from the random orientations of the wire facets relative to the electron detector. Wires exhibit diameter dependent growth directions[29] ranging from the $\langle 110 \rangle$ to the $\langle 112 \rangle$ and $\langle 111 \rangle$ for increasing wire diameters. Each growth direction exhibits different sidewall faceting.[28] In addition the appearance of sawtooth (see Fig. 12.9) structures on the NW sidewall[30] at larger diameters provides even greater variation where it was shown via UHV transmission electron microscopy that $\langle 111 \rangle$ oriented wires display predominantly $\langle 112 \rangle$ facets. Faceting may also be responsible for variations in the spectral shapes we have observed from wire to wire; the growth process randomly orients the NWs relative to the detector (see Fig. 12.5).

To compare with work functions for H-terminated planar Si surfaces we measured Φ for H-terminated planar Si surfaces in a separate higher energy (26.35 eV photons) photoemission spectroscopy system, described elsewhere,[31] and found $\Phi = 4.48$ eV, 4.42 eV and 4.56 eV for the H-terminated $\langle 110 \rangle$, $\langle 111 \rangle$ and $\langle 100 \rangle$ planar Si surfaces respectively. For the H-terminated planar Ge $\langle 111 \rangle$ surface we found $\Phi = 4.84$ eV, which is 300–400 meV larger than that for the planar H:Si surfaces. It is worth noting that the NW Φ's are smaller than the planar H:Si and H:Ge

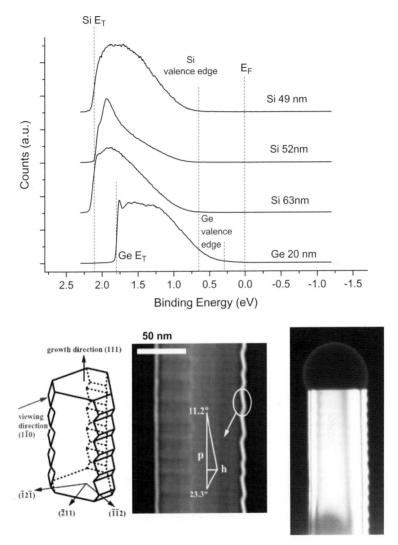

Fig. 12.9 (*Top*) Four photoemission spectra showing the variations in spectra from wire to wire. As discussed in the text, the larger NWs display faceting which gives rise to these variations due to random variations in NW structure and surfaces that face the analyzer. (*Bottom*) TEM images of 50 nm Si NW shown in cross-section revealing faceted and saw-tooth wall structure. Dark hemisphere on right is Au eutectic (Ref. 30).

values; this is not unexpected since edges associated with the faceted surfaces are expected to reduce Φ in a manner similar to that found for stepped or roughened surfaces.[32]

12.3.2.   Light Scattering from Nanowires

A strong dependence of the photoemission intensity on the polarization of the 6.2 eV relative to the wire orientation was observed (Fig. 12.8 top). Large diameter NWs ($\geq$100 nm) displayed overall greater spectral intensity when the 6.2 eV light was polarized perpendicular to the NW axis, while this effect was inverted for small diameter NWs ($\leq$ 100 nm), i.e. greater spectral intensity was observed for light polarized parallel to the NW axis. This difference is a result of the polarization dependent optical absorption and scattering properties of the wire at 6.2 eV as described by Mie theory.

To quantify this effect more thoroughly, we studied the dependence of the electron emission intensity as a function of the 200 nm light polarization for NW diameters ranging from 50 nm to 400 nm. The results are shown in Fig. 12.10 where the experimentally determined polarization anisotropies $\rho_{exp} = (I_{//} - I_{\perp})/(I_{//} + I_{\perp})$ are plotted (solid squares and circles) as a function of NW diameter. $I_{//}$ and $I_{\perp}$ are the photoemission spectral intensities collected with parallel or perpendicularly polarized light. In addition to repeatedly oxidizing and etching of an individual wire to achieve thinning (circles in Fig. 12.10) we also plot as-grown wires (squares) with a variety of diameters to insure that systematic errors associated with oxide etching did not affect the results. The insets in Fig. 12.10 show typical spectra for larger and smaller diameter wires.

In order to understand the striking inversion in the polarization dependence with NW diameter, we carried out calculations derived from Mie theory (solid curve, Fig. 12.10) for 200 nm light incident upon a Si cylinder.[33] Mie theory has been applied to the study of enhanced Raman scattering from Si NWs at visible wavelengths[34]; in the present case we investigated its applicability to conditions of strong absorption leading to photoemitted electrons with deep UV photons.

Mie theory starts with the wave equation in polar coordinates as shown below. $\mathbf{Q_{abs}}$, $\mathbf{Q_{sca}}$, and $\mathbf{Q_{ext}}$ are the absorption, scattering and

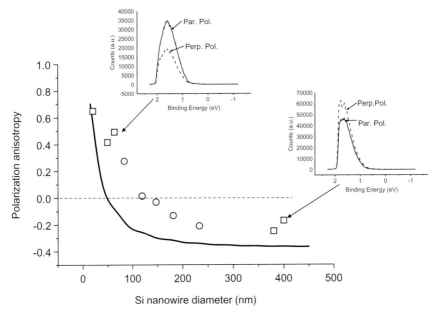

Fig. 12.10   Plot of the polarization anisotropy, ρ, of a Si NW thinned by repeated oxidation and etching in HF vapor (circles) and as-grown NWs (squares). Solid curve is the Mie theory result for the polarization anisotropy ρ as a function of NW diameter. Insets display representative spectra for large and small diameter Si NWs.

extinction efficiencies, where $\mathbf{Q_{ext}}$ is the sum of the absorption and scattering. The polarization anisotropy, determined experimentally by intensities is related to calculation through these efficiencies as shown below. At a photon energy of 6.2 eV Si is a strong absorber, with a bulk absorption coefficient (16) of $\sim$2.9 and a corresponding penetration depth of $\sim$10 nm. We applied Mie theory corresponding to the solution of the Maxwell equations for a right circular cylinder with appropriate boundary condition.[33] In this manner, the extinction $(\mathbf{Q_{ext}})$, scattering $(\mathbf{Q_{sca}})$, and absorption $(\mathbf{Q_{abs}} = \mathbf{Q_{ext}} - \mathbf{Q_{sca}})$ efficiencies for parallel and perpendicular polarization along with the corresponding polarization anisotropy $\rho_{calc} = (Q_{abs//} - Q_{abs\perp})/(Q_{abs//} + Q_{abs\perp})$ are calculated using the dielectric functions of bulk Si at 200 nm.

Following solution of the wave equation, which is done elegantly in Bohren and Huffman's exceptional text[33] on scattering of light by small particles (Fig. 12.11), the comparison between experiment and

$$\frac{1}{r}\frac{\partial}{\partial r}\frac{(r\partial\Psi)}{\partial r} + \frac{1}{r^2}\frac{\partial^2\Psi}{\partial\phi^2} + \frac{\partial^2\Psi}{\partial z^2} + k^2\Psi = 0$$

$$Q_{abs} = Q_{ext} - Q_{sca}$$

$$Q = W/DL \qquad \rho_{exp} = \frac{I_{//} - I_{\perp}}{I_{//} + I_{\perp}} \qquad \rho_{calc} = \frac{Q_{abs//} - Q_{abs\perp}}{Q_{abs//} + Q_{abs\perp}}$$

Case I: Incident light is polarized to the cylinder axis:

$$Q_{ext,I} = \frac{2}{x}\,\mathrm{Re}\left\{b_0 + 2\sum_{n=1}^{\infty} b_n\right\}$$

$$Q_{sca,I} = \frac{2}{x}\left[|b_0|^2 + 2\sum_{n=1}^{\infty}|b_n|^2\right]$$

Case II: Incident light polarized perpendicular to the cylinder axis:

$$Q_{ext,II} = \frac{2}{x}\,\mathrm{Re}\left\{a_0 + 2\sum_{n=1}^{\infty} a_n\right\}$$

$$Q_{sca,II} = \frac{2}{x}\left[|a_0|^2 + 2\sum_{n=1}^{\infty}|a_n|^2\right]$$

$$a_n = \frac{mJ_n'(x)J_n(mx) - J_n(x)J_n'(mx)}{mJ_n(mx)H_n^{(1)'}(x) - J_n'(mx)H_n^{(1)}(x)}$$

$$b_n = \frac{J_n(mx)J_n'(x) - mJ_n'(mx)J_n(x)}{J_n(mx)H_n^{(1)'}(x) - mJ_n'(mx)H_n^{(1)}(x)}$$

$$m = \frac{N_1}{N} \qquad x = \frac{\pi ND}{\lambda} \qquad N_1 = \text{NW index}, \ x = \text{size parameter}$$

Fig. 12.11 Essential components of Mie scattering calculation. The wave equation is solved in cylindrical coordinates. Q_{abs} and Q_{sca} refer to the efficiencies absorption and scattering, the sum of which is the total extinction efficiency. Two cases are analyzed, the first is where the 200 nm light polarization is aligned along the long axis of the NW and the second is for polarization perpendicular to the NW.

calculation can be carried out by considering the two cases where polarization of the 200 nm light is aligned either parallel (case I) or perpendicular (case II) to the long axis of the NW as shown above. In each case the efficiencies for scattering and extinction can be calculated through evaluation of the coefficients, a_n and b_n, comprised of Bessel and Hankel functions and their derivatives. The results of these calculations are shown in Fig. 12.10 and compared with experiment. There are no adjustable parameters in the calculation. The solid curve shown in Fig. 12.10 is a plot of ρ_{calc}.

While good qualitative agreement was obtained between theory and experimental data, the calculated values are shifted to narrower wire diameters. The inversion point is theoretically predicted at a NW diameter of $\sim$50 nm while experimentally it is found at $\sim$100 nm. Mie theory assumes an abrupt boundary between the dielectric (NW) and the surrounding medium and does not account for the detailed electronic structure and deviation of the NW from cylindrical shape (faceting). In addition, the NW surface is hydrogen-terminated. Given these sources of non-ideality, the agreement between calculation and our data to within a factor of 2 is quite reasonable and provides a framework for understanding the interaction of short wavelength light with semiconductor NWs.

12.4. DOPING OF SINGLE Si NWs

Doping of NWs is a key step in utilizing them in actual device structures. It would appear at first that doping could be easily achieved but many of the presently used approaches are problematic.[35–37] In typical integrated circuit fabrication, ion implantation is used to achieve highly accurate doping profiles and densities. Implantation into a 20 nm or smaller diameter NW would result in sputtering and damage to the crystalline structure that may not be recovered during thermal treatment. A number of workers have included dopant bearing gases such as diborane[38] (p) or phosphene[39] (n) during growth of the NW itself which does lead to incorporation, but since cracking of the dopant gases typically requires higher temperatures ($>$500°C) than those used for NW growth (430–450°C) considerable lateral growth results in larger diameter NWs. Control of the NW growth must then proceed independent of doping in order to achieve both the desired NW diameter as well as doping densities. Furthermore, it is not a given that incorporation of the dopant during growth leads to complete electrical activation; on the contrary, we will describe Al dopant incorporation during NW growth that achieves only partial electrical activation.

Measurement of the density of active dopants in a single NW is a challenge. Typically, the density of active dopants in a planar device is measured electrically, particularly since the range of dopants $(10^{14}-10^{20}/\text{cm}^3)$ is difficult to measure directly. But electrical measurement[32] of the doping density in a nm scale NW can be inaccurate due to problems with metal contacts to individual NWs. One approach to overcome this issue is to exploit the relationship between doping density and Fermi level.

For p-type doping the energetic separation between the valence band maximum (E_V) and the Fermi level can determined from the following equation[24]:

$$N_A = N_V e^{-[E_f - E_v]/KT} \tag{12.3}$$

and

$$N_V = 2[2\pi M_h^* kT/h^2]^{3/2} \tag{12.4}$$

where N_A is, for this case an acceptor (p-type) doping density, N_V is the effective density of valence band states, M_h^* is the hole effective mass for the specific semiconductor, $E_f - E_V$ is the energetic separation between the valence band maximum and the Fermi level, k is the Boltzmann constant, and T is temperature in Kelvin. To measure the doping level for an n-doped NW we have similarly:

$$N_D = N_C e^{-[E_c - E_f]/KT} \tag{12.5}$$

and

$$N_C = 2[2\pi M_e^* kT/h^2]^{3/2} \tag{12.6}$$

where N_D is the donor (n-type) doping density and N_C is the effective conduction band density of states for the specific semiconductor. Accurate measurement of the valence edge and the Fermi level will provide the needed information to determine p-type doping. For n-type doping, accurate knowledge of the band gap of the semiconducting NW is additionally needed to determine doping level. Using photoemission spectroscopy to study single NWs can therefore be used to measure the doping densities and specific examples are described here.

12.4.1. Shell Doping of Nanowires with Boron

The first experiment we discuss involves a unique method for achieving controlled doping of the NW, shown in Fig. 12.12. Our approach here is to first grow the NW under optimal conditions for diameter control. Then to controllably p-dope the NW, a thin layer of boron doped Si is added by conventional Si CVD, wherein a partial pressure of diborane gas is included during this growth step, at elevated temperature. As shown in Fig. 12.12 the desired NW doping can be achieved through simple geometrical considerations. Once the doped shell layer is deposited, high temperature annealing can be carried out to achieve complete doping of the NW. This process was carried out and the results are shown in the series of spectra shown in Fig. 12.13.

In this experiment, a 22 nm Si NW was coated with 2.5 nm thick shell doped with B, as described above. The spectrum labeled +2.5 nm B-shell reveals the result of this step. A second peak is observed in the spectrum at lower binding energy. The double peaked structure observed reflects the emission from the inner NW which matches the energetic

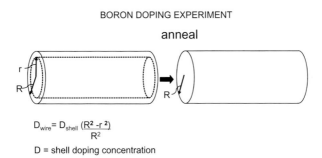

BORON DOPING EXPERIMENT

anneal

$$D_{wire} = D_{shell} \frac{(R^2 - r^2)}{R^2}$$

D = shell doping concentration

Fig. 12.12   Scheme by which controlled doping of a NW can be achieved. An undoped NW is coated with a highly doped thin layer of the same material and is subsequently annealed to both uniformly distribute the dopant atoms and to activate them in the NW matrix.

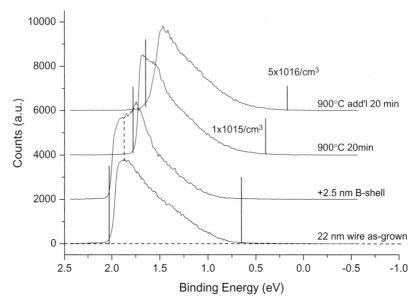

Fig. 12.13   Sequence of photoemission spectra beginning with a pristine, as-grown 22 nm NW. A thin, 2.5 nm thick shell is grown around the undoped NW. The double peaked spectrum is a result of the B-doped shell which shifts the valence edge closer to the Fermi level. The lower binding energy peak derives from the undoped inner NW core. Annealing at 900°C shifts the spectrum to lower binding energies indicating a redistribution and activation of the dopants.

position of the original NW (bottom spectrum) and the doped shell. A 20 minute anneal at 900°C shifts the spectrum to lower binding energy. Note both the valence edge as well as the threshold energy shifts and the doping density is measured to be $1 \times 10^{18}/cm^3$. The spectral shape has evolved but still reveals a second low binding energy peak. A second anneal further shifts the spectrum toward the Fermi level, the second peak is gone indicating a redistribution of dopants throughout the NW and the doping density is measured to be $5 \times 10^{18}/cm^3$.

The figure above shows the energetic shifts expected as the NW is doped revealing shifts in both the valence edge, which is used to measure the doping level via its energetic distance to the Fermi level, as well as the emission threshold. This experiment shows that the activated doping density for a B-doped NW can be accurately achieved and measured on an individual NW.

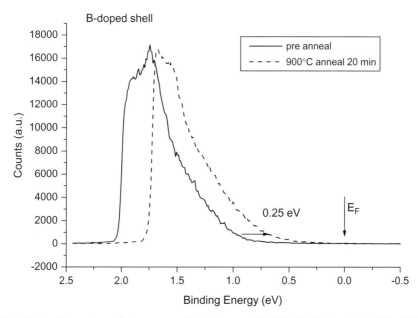

Fig. 12.14   Comparison of the pre-annealed and annealed NW with B-doped shell. The shift of the annealed spectrum to lower binding energy along with changes in the shape indicates a redistribution and electrical activation of the dopant atoms in the NW.

12.5.   AL CATALYZED NANOWIRE GROWTH AND DOPING

As described both here and in other chapters in this handbook, Au catalyzed growth has been the standard method for growing Si, Ge and myriad other semiconducting NWs via the vapor-liquid-solid phase growth mode. Although the solid solubility of Au in, e.g. Si is quite low and does not strongly dope the Si NW, it's use in the fabrication of ICs is avoided because it acts as a deep level charge trap that can severely impact carrier mobility, lifetime and diffusion.

As workers have pursued NWs and other nanostructures for device applications, alternative metal catalysts have begun to be utilized for Si NW growth.[41–43] Al has been shown to be a successful catalyst metal.[44,45] Thin layers of Al, typically on the order of 3–7 nm, are deposited on a Si substrate. Si NW growth is strongly dependent on

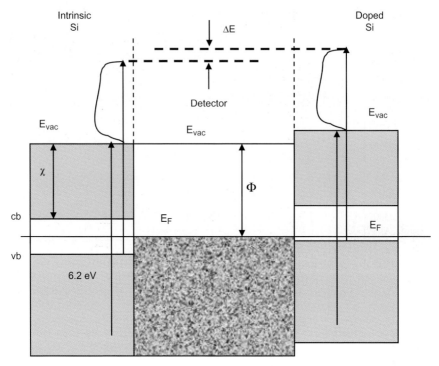

Fig. 12.15 Schematic diagram of the energetics associated with the doping of the NW. χ is the electron affinity, Φ is the work function of the detector and ΔE is the energy shift associated with the introduction of the activated dopant.

the pre-growth annealing protocol. As was found previously,[45] a pre-growth anneal at a temperature above the bulk eutectic temperature is necessary for NW growth. The Al layer is expected to agglomerate and form islands near the eutectic temperature[46] and then form Al/Si liquid droplets above the eutectic temperature.[45] For appropriate partial pressures of SiH_4 and temperatures (Fig. 12.16) Si NWs of good crystalline quality are observed to grow. Details of the Al catalyzed growth are found in the literature.

Importantly for our discussion here, Al has a substantially higher solid solubility in Si than Au and is often used as a p-type dopant in Si. As a result, following Al catalyzed growth, it is important to measure the incorporation of Al into the NW and the degree of electrical activation as a function of both growth temperature and post-growth anneal.

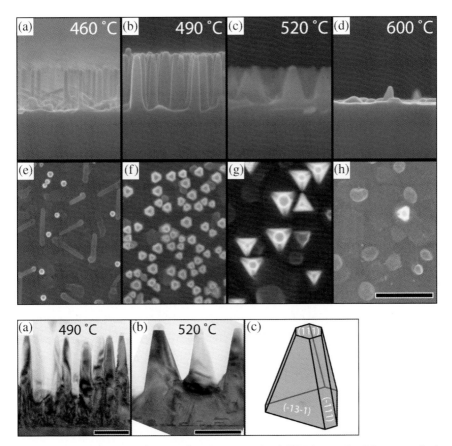

Fig. 12.16   Cross sectional (*Top*, a–d) and plan view (e–h) SEM images for different nominal NW growth temperature (indicated). For the cross sectional images the crystallographic viewing direction is varied with (a) and (c) viewed from a 11-2 direction and (b) and (d) viewed from a 1–10 direction. The scale bar, applicable to all images, is 500 nm. *Bottom panel*: (a, b) TEM of detailed NW cross-sectional structure. The structure of the NWs grown at 520°C and the faceted tops of those grown at 490°C, truncated trigonal pyramids, determined from these and similar images (from Ref. 19).

Doping of NWs and incorporation of the collector metal into the wires is of great importance to devices such as solar cells which rely heavily on controlled doping and low impurity levels. Although there is a significant solid solubility of Si in Al $(\sim 2\%)$[45] the reverse is not true. However the Al solubility in Si is not negligible in terms of doping concentration which will be important in the current study because Al is a p-type dopant in Si.[47] In order to evaluate the incorporation and

activation of Al into the Si during growth, we carried out photoelectron spectroscopy on isolated, individual NWs. Following growth, the Al
on the Si surface and at the tip of the wire was removed by exposure
to dilute 10:1 H_2O:HF solution.[48] Single isolated NWs were suspended
across open slots in a Si substrate and placed in a vacuum analysis
chamber after hydrogen termination by exposure to HF vapor in order
to eliminate surface charging. In order to evaluate the level of activated
Al within the NW we measured the location of the Si NW valence band
maximum relative to the Fermi level (located at the energy zero). Photoelectron spectra are shown in Fig. 12.17, where we compare NWs
grown with Au nanoparticles, which are nominally undoped, with those
grown using Al at various growth temperatures and anneal conditions.
The wires investigated had diameters ranging from 40 nm–120 nm. We
see that the spectra from the Al-grown NWs are shifted to lower binding energies (equivalent to the Fermi level locating closer to the valence
edge), indicating p-type doping of the NWs. Growth at 460°C resulted
in a shift of ∼100 meV to lower binding energy, while growth at 490°C
produces a 200 meV shift relative to the Au NW. These results indicate that Al is incorporated into the NW matrix and some fraction is
activated, less than $5 \times 10^{14}/cm^3$ (the active doping level calculated
from the Fermi level shift as described earlier), with higher incorporation and/or activation occurring, not surprisingly, at the higher growth
temperature.

In order to compare these Al incorporation levels with other relevant levels we consider the bulk solid phase solubility and as previously
suggested[45] incorporation of Al during aluminum induced crystalliza-
tion (AIC) at similar temperatures. The bulk solid solubility of Al in Si
has been studied for temperatures higher than and near the eutectic,[49]
but this method does not lend itself to studying the solubility below
the eutectic. We have therefore extrapolated this solid solubility study
to lower temperature (see Fig. 12.17). The active doping levels measured as grown are lower by several orders of magnitude than would
be expected by a simple extrapolation of the solid solubility. This is
also low compared to Hall-effect doping concentration measurements
for AIC at similar temperatures (the lower square point in Fig. 12.17).[50]
However it was found for AIC that not all of the Al, as measured by
secondary ion mass spectroscopy (SIMS) (the higher square point in

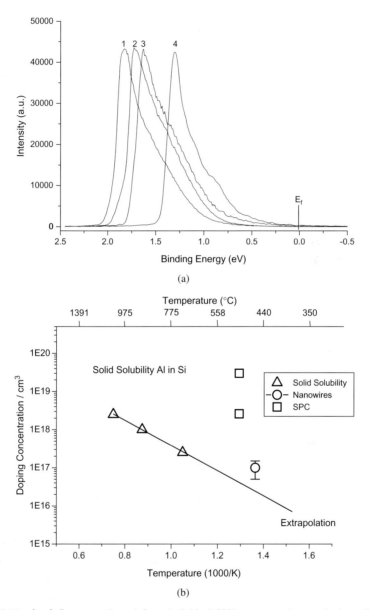

Fig. 12.17 (top) Spectra collected from individual NWs grown under nominal conditions with different metals and temperatures. E_F is the Fermi level of the system located at 0.0 eV. 1: Au growth, 2: Al growth at 460°C, 3: Al growth at 490°C, and 4: Al growth at 460°C followed by a RTA at 900°C for 10 sec. (Bottom) Triangles: solid solubility of Al in Si[49] extrapolated down through the eutectic temperature. Squares: dopant levels measured for AIC at 500°C.[26,27] Lower square: active doping level. Upper square: total incorporation. Circle: NW doping level, following a 900°C RTA, extracted from spectrum 4.

Fig. 12.17),[51] was electrically active as grown. The non-active Al could however be activated by a high temperature (900°C) anneal after which the active doping level measurements were consistent with the SIMS measurements.[51,52] We have therefore annealed the NWs in a rapid thermal anneal (RTA) process and note a substantial shift of the spectrum to lower binding energy and calculate an activated doping level of $1 \pm 0.5 \times 10^{17}/\text{cm}^3$ following the RTA (spectrum 4 Fig. 12.17 and circle). This indicates that there is more Al in these NWs than is activated during growth and indicates the actual fractional incorporation of Al in the NWs.[51]

12.6. CONTROL OF THE ELECTRONIC PROPERTIES OF Si NWs USING FUNCTIONAL MOLECULAR GROUPS

Controllably tuning the electronic properties of a semiconductor NW through the attachment of functional molecular groups to its surface is a compelling goal. The desire to control fundamental electronic properties such as valence band (VB) and conduction band (CB) electronic structure, work function Φ, and conductivity has attracted significant attention, particularly in the areas of electronic switches and sensors.[53] Given the vast variety of potential molecular modifications, it may be possible to precisely control the very electronic details that govern current flow in a transistor, light absorption in a photovoltaic cell or environmental specificity of a sensor.[54–56] In contrast to planar surfaces,[53–56] far less is known about the molecular impact to electronic properties of nanoscale structures such as NWs. Here, we develop a comprehensive picture of the interaction of a simple polar molecule, terpyridine (TP), with a Si NW through a combination of two-terminal conductivity measurements (shown schematically in Fig. 12.18), a unique single-NW photoelectron spectroscopy and theoretical analysis. Our experiments illuminate the details of molecule-NW bonding and charge transfer, diameter dependent Fermi level shifts and acid attachment leading to conductivity in the cylindrical molecular nanolayer surrounding the NW.

Our discussion begins with two-terminal conductivity measurements on Si NWs that were lithographically patterned and then trimmed to

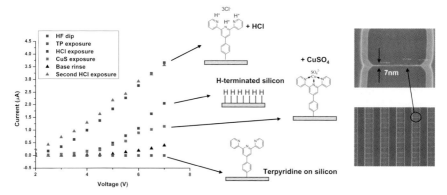

Fig. 12.18   Current vs. source/drain voltage for the lithographically patterned NWs under various termination conditions. The solid squares (red, green, blue) represent initial measurements of H-, TP- and TP + HCl-terminated NWs, followed by a base rinse in NH_3 (solid triangles) which shows the reversibility of the functionalization. Solid olive squares are for $CuSO_4$ + TP-terminated NWs. The various NW terminations are shown schematically. Scanning electron micrographs of an array of lithographically patterned NWs used for the two-terminal NWs are shown.

diameters below 10 nm via oxidation and etching. We measured the surface functionalization effects on arrays of 100–1000 patterned and suspended SiNWs, shown in Fig. 12.18. The wires were 400 nm in length and lightly p-doped (by diffusion) following annealing of the sample at 980°C for 5 seconds in N_2, required to activate the heavily p-doped contact area outside the suspended wires. We estimate doping in the contact pad area to be ~10^{18}/cm^3. We first performed measurements on hydrogen (H)-terminated NWs, followed by TP attachment and finally exposure to hydrochloric acid (TP + HCl). TP was chosen to functionalize the NW because of the presence of three basic nitrogens that could interact with a charge-bearing moiety by exposure to acids such as HCl to form amine salts, or transition metals (by coordination to metals). As such, TP represents a useful testbed to study charge transfer between an adsorbed molecular layer and a Si NW.

Figure 12.18 displays the conductivity behavior for Si NWs where the surfaces have been terminated with H (red squares), TP (green squares) and TP + HCl (blue squares); the insets display the actual termination of the NW for each condition. We note that TP termination produces a dramatic drop in conductivity relative to the H-terminated NWs. Following exposure to HCl the conductivity recovers and actually exceeds the H-terminated levels. Rinsing the NWs in a base, in this case NH$_4$OH, removes only the HCl and the conductivity returns to the original TP-terminated level (black triangles). We can then repeat the HCl exposure again (magenta triangles) reproducing the original TP + HCl results. Finally we show data from the TP-terminated NW which was exposed to CuSO$_4$ to form a coordination complex (olive squares). Most notably, the current is substantially lower than that for the HCl case, and will be discussed below.

In our photoemission experiments NWs were first H-terminated[18,28] by exposure to a vapor of hydrofluoric acid and then inserted into the analysis chamber. The NWs were then heated to 200°C in vacuum to remove any weakly adsorbed contaminants, such as atmospheric water. Each individual NW was measured and carefully followed after each exposure as will be described. Following measurement of the H-terminated NWs, the substrate was removed from the analysis system. The NW surface was then TP-terminated by exposing the substrate to a mist of terpyridylphenyldiazonium salts dissolved in acetonitrile followed by wash cycles with vapors of acetonitriles and chloroform successively, leaving a single mono-molecular layer.[58,59]

Introduction of the molecule via a mist was necessary because direct dipping of the NW covered substrates into solution washed the NWs away, presumably due to capillary forces from the liquid in the slot. With this approach NWs with diameters of 20 nm and greater were isolated for study. As an example Fig. 12.19 displays the sequence of spectra collected from a 22 nm Si NW, first H-terminated, then TP terminated and finally exposed to hydrochloric acid (HCl). The H-terminated NWs were grown undoped and exhibited a VB edge at a binding energy of 0.65 eV and an emission threshold of 2.05 eV as described before. The zero energy on the scale is the location of the Fermi energy (E_F) determined via photoemission from a clean Pt film in electrical contact to the sample.

Attachment of a monolayer of TP is observed to shift the spectrum to higher binding energy; the emission threshold is shifted by 260 meV. The direction of the spectral shift indicates that electronic charge is being transferred from the TP molecule to the Si NW; the dipole (Fig. 12.19 inset) established between the now positively charged

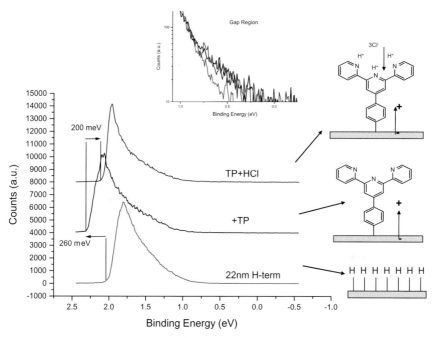

Fig. 12.19 Photoelectron spectra from H-terminated (red), TP terminated (black) and TP + HCl treated (blue), 22 nm Si NW. The spectra are vertically offset for clarity. The various terminations are shown schematically. Arrows indicate the induced dipoles. The inset displays an expanded view of the band gap region showing the additional states within the band gap resulting from the bonding of TP (and TP + HCl) with the Si NW.

molecule and the NW decreases the work function and the associated charge transfer shifts E_F upward in the gap. This simple spectroscopic observation now explains the drop in conductivity of the TP terminated NW in Fig. 12.19; the electronic charge donated from the TP electrostatically binds to the holes present from the original p-doping of the NW dramatically reducing free carriers available for conduction.

We note that both the VB edge as well as the threshold shifts to higher binding energies although their magnitudes are different. While the shift of the emission threshold is 260 meV, the VB edge moves to higher binding energy by ~200–300 meV; this indicates a shift of the Fermi level toward the conduction band. The shift of the emission threshold indicates that the work function Φ of the Si NW has decreased by 260 meV; changes in Φ provide a sensitive indication of dipole fields present at the NW surface.[60–62] For the H-terminated NW, $\Phi = 4.15\,\text{eV}$, while for the TP-terminated NW, $\Phi = 3.89$ eV. This drop in Φ is a result of the positive charge remaining on the TP which lowers the barrier for emission of the photoexcited electron. The shift to higher binding energy of the VB edge is a result of the transfer of charge from the molecule to the Si NW which shifts the Fermi level toward the conduction band.

Following monolayer coverage of TP, the sample was again removed from the analysis chamber, exposed to a vapor of HCl at room temperature for 1 minute and then inserted back into the analysis system. The emission threshold was observed to shift back toward E_F by 200 meV (Fig. 12.19, blue) resulting in $\Phi = 4.09$ eV; this can be understood by noting that the hydrogen atoms in HCl will bind with the TP nitrogen lone pair electrons; a dipole is now formed between the Cl^- and the protons attached to the TP terminal groups whose direction is opposite to that formed between the TP and the Si NW (Fig. 12.19 inset). We note that the shift of the emission threshold is not completely reversed suggesting a difference in the magnitude of the dipole moments. Furthermore, the VB edge for the TP + HCl has not recovered to the H-terminated case indicating the continued presence of a strong interaction between the adsorbed TP and the Si surface. Figure 12.20 summarizes the functionalization energetics. Over the more than 50 NWs studied in these experiments we have consistently observed these shifts, although the magnitude of the observed shifts varied by ± 50 meV for the population of NWs studied.

Additional insight into the impact on the electronic structure due to the TP monolayer can be gleaned by close examination of the NW photoemission spectra. Close inspection of the VB edges for the TP and TP + HCl spectra reveals additional states extending into the band gap region; the inset in Fig. 12.19 provides an expanded view. These states, as we will discuss further, result from bonds formed between the TP and the NW surface atoms. Because of the low intensity in this energetic region we have averaged spectra from seven different NWs for each case indicating the consistency with which excess spectral intensity is observed. This intensity is consistent with the number of TP molecules bound to the surface (we estimate the single molecular monolayer surface density to be at most $\sim 10^{14}/cm^2$) relative to the Si VB electron population accessible over the 6–10 nm electron escape depth in our

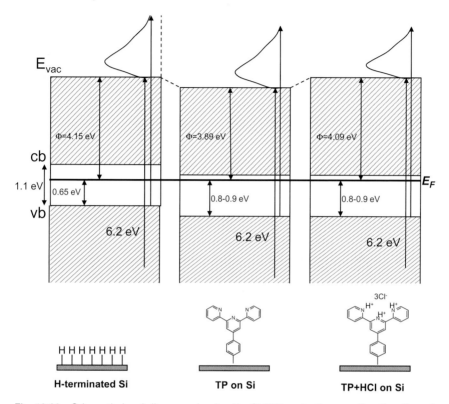

Fig. 12.20 Schematic band diagrams showing the Si NW energetics revealing the effect of the dipole fields on the work function, and band locations for various NW surface terminations.

photoemission experiments. Band gap states similar to our NW findings have been observed in planar Si/molecular systems.[55]

In order to confirm the nature of the potential shift within the dipole region, we exploited the time resolved nature of our photoemission apparatus to photoexcite the TP covered NW with a pump pulse of 800 nm (1.55 eV) light, simultaneous with the arrival of the photoemission light at 200 nm (Fig. 12.21). Photovoltage shifts have been used to measure band bending in a number of planar semiconductor systems.[31,63,64] To more completely explain the photovoltage shift we explain the nature of this effect in planar semiconductor system and we extend the approach to NWs here. Figure 12.21 displays an example of the response of a metal-insulator-semiconductor system under excitation from an intense

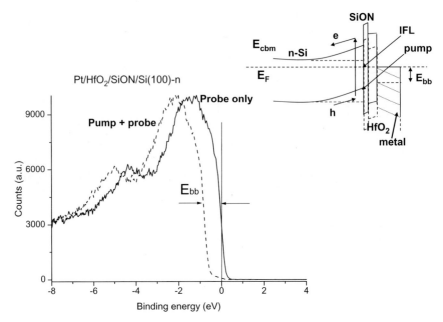

Fig. 12.21 Pump-probe photovoltage spectra for 2 nm Pt deposited on oxide covered n-Si (100). Photoexcitation with 800 nm light creates a dense e-h plasma that screens out the dipole field and flattens the bands (after Ref. 31).

pulse of 800 nm light. Initially, for a high work function metal such as Pt, deposited atop an insulator covered n-type Si substrate, electron tunneling from the Si into the metal produces a dipole field that results in an upward bend of the Si bands near the interface, as shown in Fig. 12.21.

Absorption of an intense pulse of 800 nm light generates a dense electron-hole plasma which screens the dipole field and the bands flatten. If the doping level in the Si is modest ($\sim 10^{15}/cm^3$ in this case) and/or the laser pulse is sufficiently intense, the bands can be fully flattened. This effect is observed as a rigid shift of the photoelectron spectrum to higher binding energies in the case of n-Si or lower binding energies for p-Si. Under complete band flattening conditions the band bending is extracted from the spectral shift.

The 800 nm pump creates a dense electron-hole plasma that partially screens the TP induced dipole field, shifting the VB states back toward lower binding energy (toward E_F). As can be seen in Fig. 12.22, pumping the NW shifts the spectrum by 150 meV to lower binding energy,

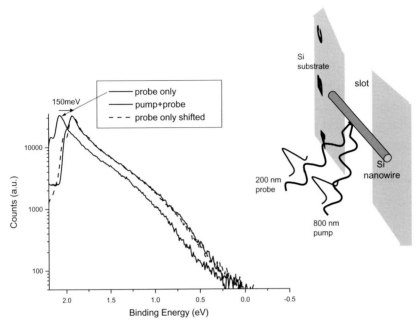

Fig. 12.22 Comparison of probe only and pump+probe spectra collected from a TP termi-nated NW. The absorption of the 800 nm pump pulse produces a dense electron-hole plasma which screens the existing dipole field. The dashed blue spectrum is that of the probe only spectrum artificially offset by 150 meV revealing a rigid shift of the pumped spectrum.

confirming the dipole nature of the shift. The dashed spectrum is the probe-only spectrum that, for comparison purposes, we have offset by 150 meV; we see that the two spectra agree, within counting statistics, indicating that the photovoltage produces a rigid electrostatic shift.

We also note that HCl functionalization increases the conductivity beyond levels measured for the H-terminated case (Fig. 12.19). Treat-ment of molecular films with acids such as HCl has been shown to sub-stantially increase hole conductivity via hopping in those systems[65,66]; such a process adds an additional conductive path, particularly if the monolayer is partially or completely ordered. From our conductivity mea-surements and assuming a single monolayer of TP + HCl, we estimate the hole mobility in the cylindrical nanolayer to be ∼0.01 cm^2/V-sec. In contrast, the complex formed with the addition of $CuSO_4$ to TP results in a smaller degree of charge transfer, a smaller dipole and lower conductivity.

The TP-induced VB shift to higher binding energy in the NW (and hence upward shift of the Fermi level), together with the elimination of p-type conductance by terpyridine adsorption, suggests a mechanism of electron donation out of the highest occupied molecular orbital (HOMO) of the adsorbed TP molecule into the Si NW. However a parabolic band-bending model, frequently used to describe charge imbalance at a surface, is inaccurate here because the NW radius (7–100 nm) is small relative to the characteristic band bending length scale in Si ($\geq 1\,\mu$m in lightly doped Si). Instead we employ a surface-local chemical model,[67] which describes the formation of a surface-TP bond, to understand the source of electron donation into the Si NW.

When the TP HOMO, assumed to have an energy level located in the semiconductor gap, forms a chemical bond with the surface, the HOMO level is shifted in energy. The surface-HOMO level coupling does however induce changes in the Si VB and CB at the surface consistent with changes observed in Fig. 12.19. The mechanism for TP doping and band edge shift derives from these two effects in a way that depends on NW radius.

12.6.1.   Theoretical Model

In our model the Si substrate is represented as a density of states $\rho(\varepsilon)$ for the VB and CB, which are composed of the 2s and 2p Si atomic orbitals most likely to be involved in bonding to the TP HOMO. The TP HOMO energy level E_a and the local Si-HOMO hopping integral V[68] are the remaining inputs into the model. A key role is played by the phase shift $\phi(\varepsilon)$ which a Si electron sees when scattering off the surface near an adsorbate atom[62]

$$\phi(\varepsilon) = \arctan\left(\frac{\Delta(\varepsilon)}{\varepsilon - E_a - \Lambda(\varepsilon)}\right), \qquad (12.7)$$

where ε is energy, $\Delta(\varepsilon) = V^2 \rho(\varepsilon)$ is the adsorbate energy level lifetime broadening, and $\Lambda(\varepsilon)$ is the Hilbert transform of $\Delta(\varepsilon)$

$$\Lambda(\varepsilon) = \frac{P}{\pi} \int_{-\infty}^{\infty} \frac{\Delta(\varepsilon')}{\varepsilon - \varepsilon'} d\varepsilon' \qquad (12.8)$$

where P is principal part. The phase shift undergoes a jump of π at the energy which solves $\varepsilon_{BS} - E_a - \Lambda(\varepsilon_{BS}) = 0$, provided that this energy lies in the Si energy gap (which occurs with typical parameter choices). The root of $\varepsilon_{BS} - E_a - \Lambda(\varepsilon_{BS}) = 0$ defines a bound state ε_{BS} in the Si energy gap to which the HOMO level is shifted. The Fermi level in the presence of a layer of adsorbate is determined by an equation for conserving particle number (the VB band is assumed fully occupied)

$$N^h_{BS} = N^e_C + N^e_S, \tag{12.9}$$

where $N^h_{BS} = 1 - f_\mu(\varepsilon_{BS})$ is the number of holes in the bound state (f_μ is the Fermi function at the adsorbate-induced chemical potential (AICP) μ), N^e_C is the change in number of electrons between the AICP and the original chemical potential μ_0

$$N^e_C = \frac{2n_S}{n_A} \int_{CB} \rho(\varepsilon)[f_\mu(\varepsilon) - f_{\mu_0}(\varepsilon)]d\varepsilon, \tag{12.10}$$

where n_S is the number of Si atoms per unit length in the NW, and n_A is the number of adsorbate atoms per unit length. The last term in Eq. (12.2) is a surface term representing the modification of the CB $\rho(\varepsilon)$ by the adsorbate

$$N^e_S = \frac{1}{\pi} \int_{CB} \phi(\varepsilon) \frac{df_\mu(\varepsilon)}{d\varepsilon} d\varepsilon. \tag{12.11}$$

In a thick NW, N^e_C dominates N^e_S. The consequence is that electrons are thermally excited from the bound state into the CB, the "doping" that positions the Fermi level between the bound state and the CB lower edge. As a result the Fermi level moves closer to the CB relative to the clean NW. Figure 12.23 shows the predicted Fermi level shift within the band gap for a range of parameters and is compared with data from the valence band shifts of individual NWs; we see that with reasonable parameters the surface-adsorbate bond model provides a good fit to the data. In Fig. 12.23, E_a is the HOMO level, with 0.0 eV corresponding to the middle of the Si gap and V is the hopping integral value.[63]

At small NW diameters the surface term, N^e_S, plays an increasingly important role. In calculating the AICP using Eqs. (12.1)–(12.4), the Si $\rho(\varepsilon)$ is modeled as two semi-elliptical bands (a conveniently soluble model[67]) mapping onto the VB and conduction bands of Si. For thin

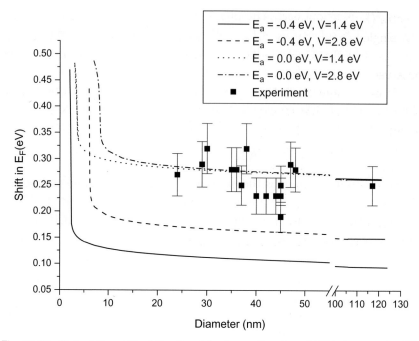

Fig. 12.23   Plot of the shift of the Fermi level as a function of NW diameter for TP-terminated NWs. The colored curves are from the theoretical treatment discussed in the text. E_a is the location of the HOMO within the NW band gap; 0.0 eV corresponds to the center of the Si NW gap. The hopping integral values, V, are discussed in the text.

wires the surface/volume ratio makes the adsorbate-induced modification of the DOS at the surface more important than thermal excitation from the HOMO as a source of electrons in the CB. This novel effect becomes very large in ultrathin wires, as seen in Fig. 12.23. Although, as we mentioned, suspended NWs below 20 nm did not survive the molecular attachment process for our photoemission experiments, the model is consistent with the slight upward trend of the data from 120 nm down to 20 nm. In addition the conductivity data in Fig. 12.18 was collected on NWs below 10 nm where the model suggests more substantial changes in E_F with molecular attachment.

The calculations demonstrate that the relative upward Fermi level shift eliminates holes in the VB, explaining the drop in conductivity in the TP covered NW. The hypothesized adsorbate-induced states lying in the lower half of the gap can be seen as a distribution in the photoemission spectra (Fig. 12.19 inset). We suggest that this distribution, as opposed to a single HOMO-NW level, arises from inhomogeneity of surface bonding sites and from inter-molecular interactions which can broaden the observed spectral emission.

These observations have potentially important implications. This behavior implies that sensor response to attachment of a polar molecule might be substantially enhanced for NWs with diameters below 10 nm but the magnitude of the response would be highly sensitive to diameter variations. For NWs with diameters >20 nm, there would be little dependence of sensitivity on NW diameter. From both experiment and theory we can now compose a picture of molecular attachment to a Si NW. TP termination results in the formation of a molecule-Si surface bond and electronic charge transfer which "n-dopes" the NW surface, pushing the Fermi level higher in the band gap by ~200–300 meV. Theory and experiment show that for NWs with diameters of 20–100 nm, only a weak dependence on diameter exists. Below 10 nm, where our conductivity measurements were made, theory predicts an increased shift of the Fermi level toward the CB. TP attachment creates a dipole field which lowers the NW work function; this is reversed with the addition of HCl which produces an opposing dipole field, but because it is spatially removed from the NW surface does not substantially modify the HOMO-NW surface bond. We suggest that the formation of a TP-HCl salt forms a conductive cylindrical nanolayer that contributes to the enhanced conductivity we observe.

12.7.   CARBON NANOTUBES

Carbon nanotubes (CNT) are essentially rolled-up cylinders of graphene that display a one-dimensional density of states.[69] These unique structures display either semiconducting or metallic properties depending on the details of their diameter and chirality. A wide range of applications have been demonstrated including electrical devices,[70] transparent conductors[71] and as additives to strengthen materials, to name just a few. While many of the properties of CNTs suggested great promise for replacing Si based devices for integrated circuits, separation of the naturally occurring mix of semiconducting and metallic species, choosing

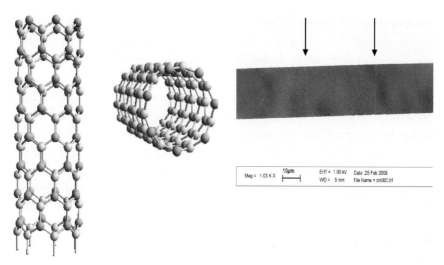

Fig. 12.24   Schematic models of a single wall carbon nanotube. Right shows SEM of 2 CNTs suspended across 15 micron gap in Si substrate.

specific diameter and chirality, and the inert nature of the surfaces of CNTs all have conspired to limit their wide-scale industrial use. Nonetheless, these intriguing structures that more recently have included other carbon-based materials such as single and multi-layer graphene[72] have come under intensive scrutiny. To this end these materials have been studied extensively via a wide range of techniques, a number of which are discussed in chapter 11 of this handbook and give extensive discussion on the formation and properties of CNTs. Those chapters give extensive discussion on the formation and properties of CNTs. In Fig. 12.24, we display a model of a single walled CNT. At right is an SEM of 2 CNTs that were grown across the slot of a Si substrate as described earlier in the chapter.

Photoelectron spectroscopy has been carried out on ensembles of CNTs and details as to the nature of the electronic and optical properties have been elucidated. In particular, the nature of the electron gas, Fermi–Dirac vs. Luttinger,[73] was addressed in a paper by Ishii et al.,[74] where close inspection of the valence edge revealed a shape that was well fit by a power law distribution consistent with the Luttinger model. In addition, the authors were able to observe van-Hove singularities which are peaks in the one-dimensional density of states and are observed as sharp

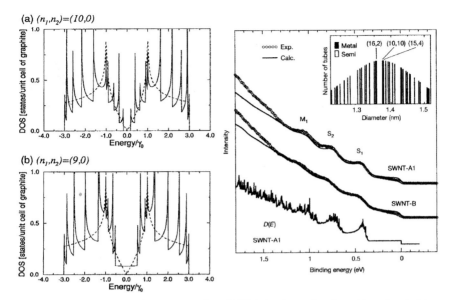

Fig. 12.25 *Left*: Density of states of single wall CNT showing series of van Hove singularities. The (10, 0) CNT is semiconducting, showing a gap in the DOS near 0 eV while the (9, 0) depicts the non-zero DOS of a metallic CNT near 0 eV (from Ref. 69) *Right*: photoemission spectra of an ensemble of CNTs on a surface showing the theoretically calculated DOS (bottom) and the experimentally collected spectra labeled SWNT-B, A1 (after Ref. 74).

features in the photoelectron spectrum broadened by the distribution of diameters in the ensemble of CNTs present on the surface of the sample. Figure 12.25 shows a theoretical calculation of the one-dimensional density of states for a semiconducting and metallic CNT (left) from the text by Saito and Dresselhaus,[69] while the right figure shows photoelectron spectra from Ishii.

In order to observe more detail in the density of states, photoelectron spectra were collected from a single, 1.8 nm CNT suspended across a gap in the Si substrate similar to that discussed previously in this chapter. The SEM picture in Fig. 12.24 shows two suspended CNTs. The experiment was carried out at room temperature in the same manner as that for single NWs of Si or Ge. The spectra for polarization parallel or perpendicular to the CNT long axis are shown in Fig. 12.26. Given the exceptionally small dimensions of the CNT studied, the count rates were quite low but high quality spectra were acquired. We note 2 peaks in the spectra derived from the van-Hove singularities as observed in

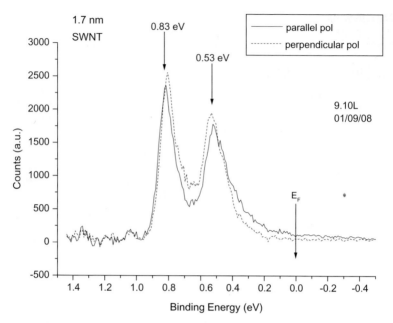

Fig. 12.26   Photoelectron spectra of an individual 1.8 nm single wall CNT taken with light polarized parallel and perpendicular to the long axis of the CNT. The peaks at 0.53 eV and 0.83 eV are van Hove singularities of the CNT. The lack of polarization sensitivity is due to the existence of the strong effect of surface plasmons on the dielectric properties of the CNT.

Fig. 12.25. Raman spectroscopy on this CNT revealed it to be single walled and semiconducting (see Freitag and Tsang). A weak decay of the density of states near the Fermi level is observed as well.

Interestingly, no difference in the spectral intensities or shapes was observed when the spectra were collected with parallel or perpendicular polarization. While for such small diameter semiconducting structures, Mie theory would suggest a large contrast, the lack of such contrast can be found in the optical response of the CNT at and near 6.2 eV photon energies. Murakami et al.,[75] have shown that while substantial anisotropy exists in the absorption coefficient below 4 eV, at ~6 eV the anisotropy collapses (see Fig. 12.27) due to surface plasmon induced changes in the electronic structure.

Clearly then we can demonstrate, at the single CNT level important electronic properties of these one-dimensional structures. As discussed in earlier sections, chemical functionalization for doping, modifying

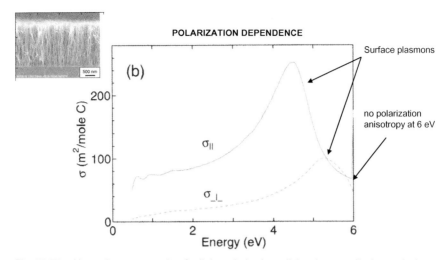

Fig. 12.27 Absorption cross-section for light polarized parallel and perpendicular to the long axis of the CNT as a function of photon energy. Near 6 eV, the cross-sections are nearly equal resulting in essentially no polarization anisotropy explaining the spectra of Fig. 12.25 (after Ref. 75).

properties to allow for selection of narrow size distributions or for patterning of CNTs essential for eventual use in integrated circuitry will all affect CNT electronic properties and will require study on the single as well as ensemble level for full understanding.

12.8. ZINC OXIDE NWS

12.8.1. Basics

Zinc oxide (ZnO) NWs have attracted considerable attention recently, owing to their unique suite of properties that include a large band gap (>3 eV), and optical transparency in the visible regime. ZnO NWs have shown promise in a myriad of applications from ultraviolet (UV) nanolasers,[76] solar cell anodes,[77] photodetectors,[78,79] to active channels in transparent flexible electronics.[80] Improving the performance of such technologies will ultimately require measurement and optimization of fundamental electronic properties such as the work function and location of the Fermi level. However, electrical assessment of such properties has been prohibited by the ability to generate quality contacts to

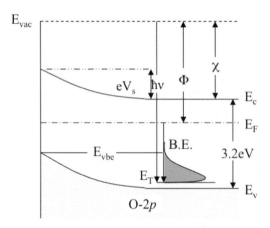

Fig. 12.28   Relation between electron energy distribution and energy levels in ZnO. Photons of energy ($\hbar v$) are absorbed and eject photoelectrons above the vacuum level (E_{vac}). The binding energies (B.E.) correspond to discrete energy levels in the semiconductor.

ZnO. A general optical diagnostic technique that can noninvasively measure such electronic properties on individual ZnO NWs is needed. Here, we discuss the application of photoelectron spectroscopy to measure the work function and Fermi level location within the ZnO NW band gap, both clean and "engineered" through the adsorption of polar self-assembled monolayers (SAMs).

    Figure 12.28 shows the energy-level diagram of a ZnO sample and how the measured photoemitted electrons relate to binding energies within the sample. It is interesting to note that the valence band of ZnO is oxygen-derived and thus, highly surface sensitive to various environments. This feature of ZnO enabled us to couple synthetic studies of different environments and various adsorbed surface species to elucidate their respected influences on surface-sensitive properties.

    ZnO NWs were prepared using a solution-based synthesis where NW arrays were hydrothermally grown from preformed "c-axis" textured nanocrystals[81] prepared on a Si substrate. The NWs have diameters ranging from 35 to 300 nm and variable lengths from 6 to 12 μm. The NWs were drop-cast onto lithographically patterned Si wafers, and individual NWs of interest were picked up and aligned by nanomanipulation (Zyvex S100). Each NW was then welded to the Si substrate by electron-beam-induced deposition of platinum and hanging over the slit by at least 4–5 microns. The Si substrates were mounted and interrogated as previously described.

Initial experiments were carried out on as-grown ZnO NWs. The position of the Fermi level was located at 0 eV binding energy, and the photoelectron emission threshold (E_T) was identified at 1.57 eV. The work function (Φ) was deduced from E_T via Eq. (12.12):

$$\Phi = h\nu - E_T \tag{12.12}$$

and for the NW in Figure 12.29, the work function is 4.63 eV. The average Φ taken on thirty different native NWs was 4.7 ± 0.2 eV,

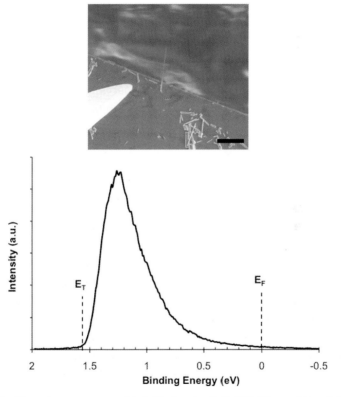

Fig. 12.29 Photoelectron spectra of individual native ZnO NW. The position of the Fermi level (E_F), which was measured from the Pt film in contact with the sample, is identified at 0 eV binding energy and the photoelectron emission threshold (E_T) is also identified at 1.57 eV. The work function (Φ) of the NW was deduced from E_T, and for this NW, the work function is calculated as 4.63 eV. Inset is SEM image of ZnO NW which spectra was taken on, partially extending into lithographically patterned gap. Scale bar is 5 μm.

approximately 0.5 eV lower than its bulk value. For the range of NW diameters studied (90–200 nm), a dependence on diameter was not found. The hexagonal faceting of the NW, orientation of the facet/edge relative to the detector and cleanliness of the surface all may have been responsible for the reduction in Φ. This effect, which was also seen in Si NWs, is not unexpected since edges associated with the faceted surfaces are expected to reduce Φ in a manner similar to that found for stepped or roughened surfaces.[82]

12.8.2. Oxygen Vacancies

Native ZnO is intrinsically n-type, owing to its primary defects, zinc interstitials and oxygen vacancies. The high density of surface defects on as-grown ZnO NWs can act as 'binding magnets' for adsorbed species, such as water, carbon dioxide and oxygen. Extrinsic surface states associated with dangling bonds of such adsorbates trap holes or electrons, and produce a space-charge layer.[83] The width of the space-charge layer (W) plays a critical role in the carrier dynamics of the NWs and can be calculated according to equation 12.13

$$W = \sqrt{(2\varepsilon_{ZnO}\,\Phi_s)/(eN_D)} \qquad\qquad (12.13)$$

where ε_{ZnO} is the dielectric constant of ZnO (8.66), Φ_s is the surface barrier potential or built-in potential (0.3 eV), e is the charge of an electron and N_D is the doping density of the material ($5 \times 10^{18}/cm^3$). The space charge layer or depletion width can then be calculated as 23 nm. Since the depletion width can be altered by electronegative adsorbates on the NW surface, removal should modulate the bands and depletion width. To remove water vapor and other surface species (CO_2, O_2); the substrate was annealed in-situ at 300°C for 1 minute. Figure 12.30 demonstrates spectra taken on an individual NW before and after annealing. A shift of the valence band edge towards lower binding energies was detected, which was induced primarily by thermolysis of water vapor and detachment of depletive ions bound to the surface. The shift may also be attributed to the influence of the co-adsorption of water on the adsorption of another electron acceptor adsorbate.[84] In any of these mechanisms, thermal desorption of adsorbed water and other oxidizing

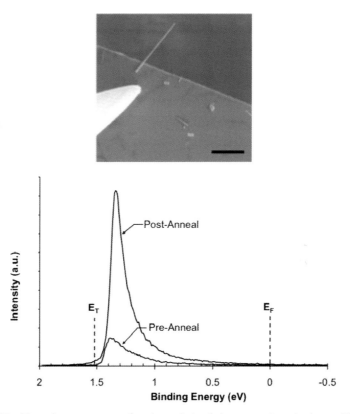

Fig. 12.30 Photoelectron spectra after thermolysis of electronegative adsorbates. The electron emission threshold ($E_T = 1.53$ eV) and Fermi level ($E_F = 0$) are labeled by dashed lines and the work function (Φ) for this NW is 4.67 eV. Inset is SEM image of NW probed (diameter = 270 nm) where the scale bar is 5 μm.

species increased the density of bulk and surface oxygen vacancies, which in ZnO act as electron donors. The increase in the number of oxygen vacancies caused electrons to become trapped in "f-center" defects and accumulate negative charge at the surface, bending the bands slightly upward.

Since the number of equilibrium surface and bulk oxygen defects in ZnO is a function of the environmental oxygen partial pressure and temperature,[85] the samples were annealed *in-situ* at 300°C for 1 minute in 10^{-5} torr O_2. Figure 12.31 demonstrates spectra taken before and after annealing an individual ZnO NW in a partial pressure of oxygen.

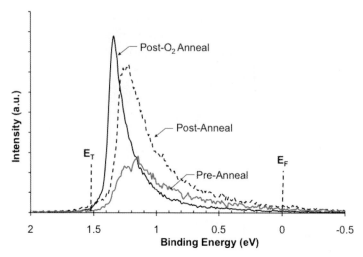

Fig. 12.31   Photoelectron spectra after annealing in dry oxygen. The spectra of a native NW, post-annealed, and annealed in a partial pressure of oxygen are shown. A somewhat smaller movement of the valence edge away from the Fermi level was detected for this particular NW. E_T was measured at 1.56 eV for this NW (diameter $= 220$ nm) and the corresponding work function is calculated to be 4.64 eV.

Upon introduction into the chamber, adsorbed oxygen molecules repopulated the surface vacancies, became ionized and induced a spectral shift away from the Fermi level. Large changes to the surface potential due to small changes in the partial pressure of oxygen are substantiated by the fact that the states near the Fermi level are mostly oxygen derived.[86] Carbon in and on the NW surface may have also been responsible for the energetic shift; as oxygen is chemisorbed onto the surface and an electron is captured, a depletive CO_2^- ion could have been formed. The decrease of surface oxygen vacancies effectively abated the number of filled (donor) intragap states, which in turn reduced the surface charge and band bending.

To reverse the effect of adsorbed oxygen, NWs were irradiated with UV light for an extended period. Upon absorption of optical pulses, freed electrons can then partake in a variety of processes such as scattering into surface states (capture). The photogenerated holes can migrate to the surface along the potential slope created by band bending[87] and combine with surface state electrons, excising the electronegative surface species (O_2^- and CO_2^-). Figure 12.32 displays the photoemission

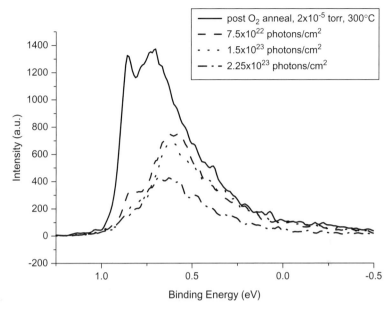

Fig. 12.32 Prolonged exposure of ZnO NW to UV light induces the photoemission threshold (E_T) to shift towards the Fermi level, indicating upwards band bending, and photolysis of O_2^- and CO_2^- from the NW surface. The top curve is a spectrum of a native NW and the accumulated photon flux associated with each spectrum taken thereafter is $7.5e^{22}$ photons/cm^2, $1.5e^{23}$ photons/cm^2, and $2.25e^{23}$ photons/cm^2.

threshold (E_T) shift towards the Fermi level with extended exposure. This result was qualitatively consistent with generation of surface oxygen vacancies by annealing, whereby donated electrons get trapped in "f-center" defects and acceptor molecules are discharged. The localized charge layer bent the bands upward again. These results illustrate the high sensitivity of the ZnO NW surface to molecular adsorption processes, which makes them ideal candidates to study other adsorption signatures such as those generated from covalently bound polar organic molecules. Adsorption of polar organic molecules on semiconductor surfaces has been shown to induce significant changes in the electrostatic potential and in turn, electron affinity (χ) and work function (Φ). To date quantifying the affect that SAMs exert on various physical properties has been challenging, as electrical contacts to passivated surfaces often degrade the organic monolayer. Photoelectron spectroscopy on individual NWs is a salient method for correlating the

interaction between adsorbed dipolar molecules and NW electronic structure because of the surface sensitivity of the technique.

12.8.3.   Surface Fuctionalization

SAMs of various phosphonic acid (PA) derivatives were chosen as a
platform to study changes in Φ because PAs are strong binders to
metal oxides[88] and have been shown to have a pronounced effect on the
electronic properties of ZnO thin films.[89] Additionally, the high density
of surface binding sites in ZnO provided excellent accessibility for PA-
based SAMs to target the surface and form up to three bonds at the
interface,[90] exerting a strong molecule-surface coupling. Moreover, the
ability to modify the para-substituents of the PAs and maintain identical
surface binding groups enabled the capability to change the NW work
function controllably.

Figure 12.33 demonstrates spectra taken before and after functionalization with benzyl phosphonic acid (BPA), an electron-donating SAM
with net dipole moment pointing towards the surface. The spectra and
E_T for a native NW (bottom spectrum) identified in Fig. 12.33 at 1.59 eV
has clearly been shifted by 390 meV to 1.20 eV after passivation with
BPA (middle spectrum), indicating a large change in the effective work
function. To test whether the effect of the adsorbed molecule could be
reversed, the substrate was annealed in-situ at 300°C for 2 minutes.
Figure 12.33 shows the spectrum shift back towards its original state
(top spectrum), but not fully recover due to residual hydrocarbon left
from the SAM. The spectra did not fully recover but proved a near
reversal of the molecular modified surface can be achieved by annealing.
Figure 12.33(b) demonstrates a representative spectrum taken before
and after functionalization with 1-octyl phosphonic acid (OPA). OPA is
also an electron-donor phosphonic acid but is an aliphatic system, and
was selected to contrast the aryl-phosphonic acid SAMs used. Alkylphos-
phonic acids have been shown to have higher molecular packing densities
and exert better coverage than aromatic SAMs. For the native NW, E_T
was identified in Fig. 12.33(b) at 1.6 eV. After functionalization, E_T was
spectroscopically shifted by 210 meV to 1.40 eV, indicating a smaller
change in the effective work function than for BPA. The results of the

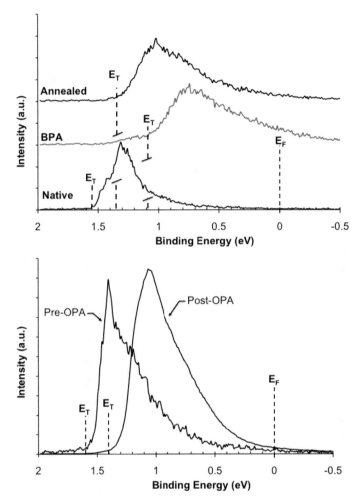

Fig. 12.33 Photoemission spectra of ZnO NWs coated with electron-donating SAMs. (a) Photoemission spectrum taken on native NW (bottom), after functionalization with benzyl phosphonic acid (middle), and post annealing the functionalized NW *in-situ* (top). The spectra have been offset for clarity. The electron emission thresholds (E_T) for each are labeled by dashed lines (bottom ~1.59 eV, middle ~1.20 eV, top ~1.36 eV). The spectra did not fully recover due to residual hydrocarbon left from the SAM. The work function is 4.61 eV for native ZnO, 5.0 eV for functionalized ZnO, and 4.84 eV for a functionalized NW that has been annealed. The diameter of the NW probed was 150 nm. (b) Photoemission spectrum taken on clean NW (left), and after functionalization with 1-octyl phosphonic acid (right). E_T was been shifted by 0.2 eV for the OPA functionalized wire and is labeled by a dashed line. The diameter of the NW probed was 170 nm.

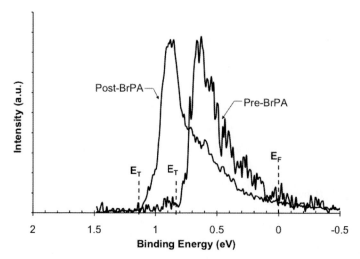

Fig. 12.34 Photoemission spectra of ZnO NWs coated with electron-accepting SAMs. 4-bromophenyl phosphonic acid (4-BrPA) is an electron-accepting molecule with dipole moment similar to BPA. The NW functionalized with 4-BrPA (right), when compared to its clean spectrum (left), showed an energetic shift of E_T, by nearly 0.4 eV away from the Fermi level. The measurement was done on a NW with diameter of 160 nm.

NWs functionalized with BPA and OPA are qualitatively consistent with the results of other electron donors at the surface (oxygen vacancies), where a shift towards the Fermi level was detected.

To gauge if the dipolar effect could be shifted in the reverse direction using an oppositely charged dipole, an aryl electron-accepting phosphonic acid was grafted onto the NW surfaces using the T-BAG method. 4-bromophenyl phosphonic acid (4-BrPA), an electron-withdrawing molecule pointing away from the surface with slightly stronger dipole moment than BPA[91] was adsorbed. Figure 12.34 shows a representative spectrum of a NW passivated with 4-BrPA. The spectra and photo-electron emission threshold for a clean NW (pre-BrPA) were shifted by 390 meV after passivation (post-BrPA), indicating a large change in the effective work function. For the number of samples treated with 4-BrPA, the average shift was 390 meV. The results were qualitatively consistent with generation of surface acceptors such as oxygen or water vapor, whereby a shift away from the Fermi level was detected. The effect of various molecules on the NW properties is shown below in Fig. 12.35 along with their dipole moments. The dipole moments were calculated

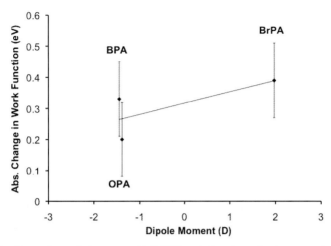

Fig. 12.35 Comparison of changes to ZnO NW work functions. As the strength of the net dipole moment increases, the change in work function also increased.

with Spartan software (Wavefunction Inc.) in an equilibrium geometry in the ground state using the Hartree-Fock method and 3-21G* basis set.

To gauge the binding mode(s) of the molecules to the NW surface, Fourier Transform Infrared (FTIR) spectroscopy in attenuated total reflectance mode (Thermo Mattson Infinity Gold) was used. Strong binding between the aryl SAMS (BrPA and BPA) and the NW surface was found, as evidenced by strong methylene asymmetric ($v_a(CH_2)$) and symmetric ($v_s(CH_2)$) stretching absorptions at 2918 and 2850 cm^{-1}, respectively (Fig. 12.36(a)). The P–O stretching region between 1300 and 800 cm^{-1} also demonstrated considerable changes in the number and frequencies of the P=O and P–O stretching bands (990–1010, 1130, 1260–1300 cm^{-1}), which is indicative of a pronounced interaction of the phosphonate headgroup with the ZnO surface (Fig. 12.36(b)). Also apparent is the disappearance of the peaks at 1200 and 950 cm^{-1}, which was also observed for phenylphosphonic acid on ZrO_2^{92} and ITO.[93] The absence of these two bands, which were assigned to the P=O and P-O-H stretching vibrations, indicate that the aryl SAMs are bound to the surface in a tridentate fashion. For the alkyl system (OPA), the methylene stretching absorptions show the binding was much less pronounced. One possible case for the lack of strong binding could be the phosphonic acids reacted with the surface to make a phosphonate salt.

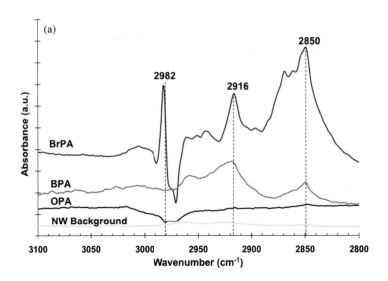

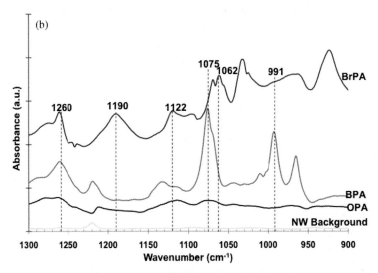

Fig. 12.36   Fourier Transform Infrared (FTIR) spectra of NWs grafted with various SAMs.
(a) The methylene stretching vibration bands (2915 and 2848 cm^{-1}) show enhanced stretching modes for the aryl SAMs (BrPA and BPA). The NWs functionalized with OPA show little
or no enhancement, indicating weak binding. (b) The changes in the number and frequencies
of the P=O and P–O stretching bands (990–1010, 1130, 1260–1300 cm^{-1}) show tri-dentate
surface bonding of the aryl SAMs (BrPA and BPA) to the NW surface. The P–O stretching
region (1300–800 cm^{-1}) of OPA also showed relatively weak interaction, which corroborates
the spectroscopic results of a relatively weak interaction between the SAM and the NW
surface.

Or another scenario could be the phosphonic acid formed a weak hydrogen bond to the surface and perhaps to neighboring molecules. The changes to the P–O stretching region also showed a relatively weak interaction, which corroborates the magnitude of the small spectroscopic shift. While the ranges for the various P-O stretching vibrations overlap and vary in their level of metal binding, conclusive assignment of the bands was precluded.

The changes in Φ due to adsorption of phosphonic acids stem from several molecular contributions. The first is a constant charge transfer effect from the phosphonic acid binding groups to the semiconductor surface that occurs from binding. The next is the effect of the electronegativity of the headgroup of the molecule. The headgroup contains the functional element that determines the net dipole moment of the entire molecule, and the direction in which it points (to or from the surface). In addition to the dipolar effect, the molecules can also change the net surface charge and depletion width by modulating the distribution of surface states.

The work function and location of the Fermi level of hydrothermally grown ZnO NWs was measured using low energy photoelectron spectroscopy. Systematic surface treatments were conducted to reveal the impact of environmental adsorbates on the work function as well as allow estimation of their effects on the surface densities of states. The technique was then extended to assess the interactions between various dipolar phosphonic acid-based SAMs and the NW work function.

12.9. CONCLUSIONS

In this chapter we have endeavored to describe the instrumentation and techniques surrounding the new area of single NW photoelectron spectroscopy. This discussion was punctuated with specific experiments designed to further elucidate the advantages of studying the electronic and optical properties of nanostructures, one at a time. One can envision capitalizing on many of these initial experiments to investigate heterostructure NWs, compound semiconductor NWs, single and multiple layered materials such as chalcogenides and graphenes with exceptional spatial resolution and more. We have suggested that higher photon energies for surface specificity and core level analysis for NW surface and bulk chemistries would yield important new information about nanostructures. Higher repetition rates for better signal averaging in ground state and photoexcited systems would permit the studies of electron dynamics on an unprecedented spatial scale.

REFERENCES

1. E. W. Plummer, W. Eberhardth, *Adv. Chem. Phys.* (Wiley, NY) 1982.
2. R. Haight, *Surf. Sci. Rep.* **21**, 275 (1995).
3. B. K. Ridley, *Quantum Processes in Semiconductors*, Oxford University Press (1988).
4. D. E. Aspnes, *Surf. Sci.* **132**, 406 (1983).
5. W. F. Brinkman, T. M. Rice, *Phys. Rev. B* **7**, 1508 (1975).
6. K. Giesen, F. Hage, F. J. Himpsel, H. J. Riess, W. Steinmann, *Phys. Rev. Lett.* **55**, 300 (1985).
7. K. Read, H. S. Karlsson, M. M. Murnane, H. C. Kapteyn, R. Haight, *J. Appl. Phys.* **90**, 294 (2001).
8. D. Lim, R. Haight, *J. Vac. Sci. Technol.* **A23**, 1698 (2005).
9. J. A. R. Samson, *Techniques of Vacuum Ultraviolet Spectroscopy*, Pied Publications (1989).
10. T. Tchin, *J. Opt. A: Pure Appl. Opt.* **5**, 73 (2003).
11. W. Chao, B. D. Harteneck, J. A. Liddle, E. H. Anderson, D. T. Atwood, *Nature* **435**, 1210 (2005).
12. See, for example, Murray and Noyan, Chapter 7, this publication.
13. E. Spiller, *Soft X-ray Optics*, SPIE Opt. Eng. Press (1994).
14. E. Bauer, M. Mundschau, W. Sweich, W. Telieps, *Ultramicroscopy* **31**, 49 (1989).
15. M. Y. Sfeir, T. Beetz, F. Wang, L. M. Huang, X. M. H. Huang, M. Y. Huang, J. Hone, J. A. Misewich, T. F. Heinz, L. J. Wu, Y. M. Zhu, L. E. Brus, *Science* **312**, 554 (2006).
16. D. E. Aspnes, A. A. Studna, *Phys. Rev. B* **27**, 985 (1983).
17. M. P. Seah, W. A. Dench, *Surf. Int. Anal.* **1**, 2 (1979).
18. R. Haight, G. Sirinakis, M. C. Reuter, *App. Phys. Lett.* 233116-1-3 (2007).
19. B. A. Wacaser, M. C. Reuter, M. M. Khayyat, C. Y. Wen, R. Haight, S. Guah, F. M. Ross, *Nano Lett.* **9**, 3296 (2009).
20. R. Haight, L. Sekaric, A. Afzali, D. Newns, *Nano Lett.* **9**, 3165 (2009).
21. C. A. Aguilar, R. Haight, A. Mavrokefalos, B. A. Korgel, S. C. Chen, *ACS Nano* **3**, 3057 (2009).
22. A. Einstein, *Ann. Phys.* (Leipzig) **8**, 149 (1905).
23. D. Cahen, A. Kahn, *Adv. Mater.* **15**, 271 (2003).
24. S. M. Sze, *Physics of Semiconductor Devices*, John Wiley & Sons (1981).
25. R. J. Jones, K. D. Moll, M. J. Thorpe, J. Ye, *Phys. Rev. Lett.* **94**, 193201 (2005).

26. R. S. Wagner, W. C. Ellis, *Appl. Phys. Lett.* **4**, 89 (1964); R. S. Wagner, in *Whisker Technology*, ed. By A.P. Levitt (Wiley Interscience, N.Y. 1970).
27. Y. F. Zhang, L. S. Liao, W. H. Chan, S. T. Lee, R. Sammynaiken, T. K. Sham, *Phys. Rev. B* **61**, 8298 (2000).
28. D. D. Ma, C. S. Lee, F. C. Au, S. Y. Tong, S. T. Lee, *Science* **299**, 1874 (2003).
29. Y. Wu, Y. Cui, L. Huynh, C. J. Barrelet, D. C. Bell, C. M. Lieber, *Nano Lett.* **4**, 433 (2004).
30. F. M. Ross, J. Tersoff, M. C. Reuter, *Phys. Rev. Lett.* **95**, 146104-1 (2005).
31. D. Lim, R. Haight, M. Copel, E. Cartier, *Appl. Phys. Lett.* **87**, 72902-1 (2005).
32. K. Bescocke, B. Krahl-Urban, H. Wagner, *Surf. Sci.* **68**, 39 (1977).
33. See for example, C. Bohren, D. R. Huffman, *Absorption and Scattering of Light by Small Particles* (Wiley) 1983, pp. 202–213, which describes calculations for absorption an scattering of light from an infinite right circular cylinder.
34. L. Cao, B. Nabet, J. E. Spanier, *Phys. Rev. Lett.* **96**, 157402 (2006).
35. S. Ingole, P. Aella, P. Manandhar, S. B. Chikkannanavar, E. A. Akhadov, D. J. Smith, S. T. Picraux, *J. Appl. Phys.* **103**, 104302-1 (2008).
36. L. Pan, K. K. Lew, J. M. Redwing, E. C. Dickey, *J. Cryst. Growth* **277**, 428 (2005).
37. L. J. Lauhon, M. S. Gudiksen, C. L. Wang, C. M. Lieber, *Nature* **420**, 57 (2002).
38. S. J. Whang, S. Lee, D. Z. Chi, W. F. Yang, B. J. Cho, Y. F. Liew, D. L. Kwong, *Nanotech.* **18**, 275302 (2007).
39. N. Fukata, M. Mitome, Y. Bando, M. Seoka, S. Matsushita, K. Murakami, J. Chen, T. Sekiguchi, *Appl. Phys. Lett.* **93**, 203106 (2008).
40. M. T. Bjork, H. Schmid, J. Knoch, H. Riel, W. Riess, *Nature Nootech.* **4**, 103 (2009).
41. A. M. Morales, C. M. Lieber, *Science* **279**, 208 (1998).
42. T. I. Kamins, R. S. Williams, Y. Chen, Y. L. Chang, Y. A. Chang, *Appl. Phys. Lett.* **76**, 562 (2000).
43. B. M. Kayes, M. A. Filler, M. C. Putnam, M. D. Kelzenberg, N. S. Lewis, H. A. Atwater, *Appl. Phys. Lett.* **91**, 103110 (2007).
44. P. Rai-Choudhary, W. T. Takei, *J. Electrochem. Soc.* **121**, 1228 (1974).
45. Y. Wang, V. Schmidt, S. Senz, U. Gosele, *Nat. Nano* **1**, 186 (2006).
46. J. J. Bart, *IEEE Trans. Elect. Dev.* **16**, 351 (1969).
47. R. Rurali, X. Cartoixal, *Nano Lett.* **9**, 975 (2009).
48. S. J. Whang, S. J. Lee, W. F. Yang, B. J. Cho, Y. F. Liew, D. L. Kwong, *Electrochem. Sol. St. Lett.* **28**(10), E11 (2007).
49. D. Navon, V. Chernyshov, *J. Appl. Phys.* **28**, 823 (1957).
50. Y. Civale, L. K. Nanver, P. Hadley, EJ. G. Goudena, H. Schellevis, *IEEE Elect. Dev. Lett.* **27**, 341 (2006).
51. O. Nast, S. R. Wenham, *J. Appl. Phys.* **88**, 124 (2000).
52. O. Nast, S. Brehme, D.-H. Neuhaus, S. R. Wenham, *IEEE Trans. Elect. Dev.* **46**, 2062 (1999).
53. F. Seker, K. Meeker, T. F. Kuech, A. B. Ellis, *Chem. Rev.* **100**, 2505 (2000).
54. A. Vilan, D. Cahen, *Trends in Biotechnology*, **20**, 22 (2002).
55. L. Segev, A. Salomon, A. Natan, D. Cahen, L. Kronik, *Phys. Rev. B* **74**, 165323 (2006).

56. B. Chen, A. K. Flatt, H. Jian, J. L. Hudson, J. M. Tour, *Chem. Mater.* **17**, 4832 (2005).
57. O. Gunawan, L. Sekaric, A. Majumdar, M. Rooks, J. Appenzeller, J. W. Sleight, S. Guha, W. Haensch, NanoLett. **8**, 1566 (2008).
58. M. P. Stewart, F. Maya, F. V. Kosynkin, S. M. Dirk, J. J. Stapleton, C. L. McGuiness, D. L. Allara, J. M. Tour, *Amer. Chem. Soc.* **126**, 370 (2004).
59. D. K. Aswal, S. P. Koiry, B. Jousselme, S. K. Gupta, S. Palacin, J. V. Yakhimi, *Physica E* **41**, 325 (2009).
60. D. Cahen, A. Kahn, *Adv. Mater.* **15**, 271 (2003).
61. T. He, H. Ding, N. Peor, M. Lu, D.A. Corley, B. Chen, Y. Ofir, Y. Gao, S. Yitzchak, J. M. Tour, *J. Amer. Chem. Soc.* **13**, 1699 (2008).
62. T. He, D. A. Corley, M. Lu, N. H. Di Spigna, J. He, D. P. Nackashi, P. D. Franzon, J. M. Tour, *J. Amer. Chem. Soc.* online at http//. pubs.acs.org/doi:10.1021/ja9002537.
63. M. Bruening, E. Moons, D. Cahen, A. Shanzer, *J. Phys. Chem.* **99**, 8368 (1995).
64. L. Kronik, Y. Shapira, Y. Surf, *Sci. Rep.* **36**, 1 (1999).
65. A. G. MacDiarmid, J. C. Chiang, A. F. Richter, A. J. Epstein, *Synth. Met.* **18**, 286 (1987).
66. M. Pope, C. E. Swenberg, *Electronic Processes in Organic Crystals and Polymers*, Oxford.
67. D. M. Newns, *Phys. Rev.* **178**, 1123 (1969).
68. R. G. Parr, *Quantum Theory of Molecular Electronic Structure*, Benjamin New York (1964).
69. R. Saito, G. Dresselhaus, M. S. Dresselhaus, *Physical Properties of Carbon Nanotubes*, Imperial College Press (1999).
70. Z. H. Zen, J. Appenseller, Y. M. Lin, J. Sippel-Oakley, A. G. Rinzler, J. Y. Tang, S. J. Wind, P. M. Solomon, P. Avouris, *Science* **311**, 1735 (2006).
71. M. A. Contreras, T. Barnes, J. van de Iagemaat, G. Rumbles, T. J. Coutts, C. Weeks, P. Glatkowski, I. Levitsky, J. Peltola, D. A. Britz, *J. Phys. Chem. C* **111**, 14045 (2007).
72. Y. M. Lin, P. Avouris, *Nano Lett.* **8**, 2119 (2008).
73. J. Voit, *Rep. Prog. Phys.* **57**, 977 (1994).
74. H. Ishii, H. Kataura, H. Shiozawa, H. Otsuto, Y. Takagama, T. Miyahara, S. Suzuki, Y. Achiba, M. Nakatake, T. Nariumara, H. Higashiguchi, K. Shimada, H. Namatame, M. Taniguchi, *Nature* **426**, 540 (2003).
75. Y. Murakami, E. Einarsson, T. Edamura, S. Maruyama, *Phys. Rev. Lett.* **94**, 087402 (2005).
76. M. Huang, S. Mao, H. Feick, H. Yan, H. Wu, H. Kind, E. Weber, R. Russo, P. Yang, *Science* **292**, 1897 (2001).
77. M. Law, L. Greene, J. Goldberger, J. Johnson, R. Saykally, P. Yang, *Nat. Mater.* **4**, 455 (2005).
78. H. Kind, H. Yan, B. Messer, M. Law, P. Yang, *Adv. Mater.* **14**, 158 (2002).
79. C. Soci, A. Zhang, B. Xiang, S. Dayeh, D. Aplin, J. Park, X. Bao, Y. Lo, D. Wang, *Nano Lett.* **7**, 1003 (2007).
80. S. Ju, A. Facchetti, Y. Xuan, J. Liu, F. Ishikawa, P. Ye, C. Zhou, T. Marks, D. Janes, *Nat. Nano.* **2**, 378 (2007).

81. L. Greene, M. Law, D. Tan, M. Montano, J. Goldberger, G. Somorjai, P. Yang, *Nano Lett.* **5**, 1231 (2005).
82. K. Bescocke, B. Krahl-Urban, H. Wagner, *Surf. Sci.* **68**, 39 (1977).
83. Y. Takahashi, M. Kanamori, A. Kondoh, H. Minoura, Y. Ohya, *Jpn. J. Appl. Phys.* **33**, 6611 (1994).
84. N. Barsan, U. Weimar, *J. Electroceram.* **7**, 143 (2001).
85. M. Arnold, P. Avouris, Z. Pan, Z. Wang, *J. Phys. Chem. B* **107**, 659 (2003).
86. K. Jacobi, G. Zwicker, A. Gutmann, *Surf. Sci.* **141**, 109 (1984).
87. A. Kolmakov, M. Moskovits, *Annu. Rev. Mater. Res.* **34**, 151 (2004).
88. W. Gao, L. Dickinson, C. Grozinger, F. Morin, L. Reven, *Langmuir* **12**, 6429 (1996).
89. A. Salomon, D. Berkovich, D. Cahen, *Appl. Phys. Lett.* **82**, 1051 (2004).
90. P. Mutin, G. Guerrero, A. Vioux, *J. Mater. Chem.* **15**, 3761 (2005).
91. D. Cahen, G. Hodes, *Adv. Mater.* **14**, 789 (2002).
92. J. Randon, P. Blanc, R. Paterson, *J. Membrane Sci.* **98**, 119 (1995).
93. S. Koh, K. Mcdonald, D. Holt, C. Dulcey, J. Chaney, P. Pehrsson, *Langmuir* **22**, 6249 (2006).

13

TIME-DOMAIN THERMOREFLECTANCE MEASUREMENTS FOR THERMAL PROPERTY CHARACTERIZATION OF NANOSTRUCTURES

Scott T. Huxtable
Department of Mechanical Engineering
Virginia Tech., Blacksburg, VA 24061-0002, USA

13.1. INTRODUCTION AND BACKGROUND

In the past 25 years, there have been several noteworthy developments regarding nanoscale thermometry.[1,2] The 3ω method[3] uses a micro-fabricated metal line as a heater and a thermometer and is capable

of accurately measuring the thermal conductivity of films as thin as
$\sim$100 nm. While this technique is one of the most widely used for thermal
conductivity measurements on thin films, the 3ω technique measures the
total thermal resistance below the metal line and cannot easily distinguish between the thermal resistance due to interfaces and the thermal resistance internal to the individual films or nanostructures. It also
requires photolithography on the sample, which can be challenging or
tedious for certain samples.

Scanning thermal microscopy[4] (SThM) has also emerged as a useful tool in nanoscale thermal characterization. The majority of recent
SThM work essentially involves fabricating a thermocouple junction at
the tip of an atomic force microscope (AFM) cantilever and allows for
the qualitative study of thermal transport phenomena with sub 50 nm
resolution. However, it is extremely difficult to quantitatively measure
temperature, or thermal conductivity, using SThM due to uncertainties
regarding tip-sample thermal contact and heat transfer.

Another notable advance has come in the area of high-speed
lasers where non-contact optical approaches based on thermoreflectance
have been developed.[5-7] The thermoreflectance techniques typically use
a high-speed laser to produce pulses that are split into "pump" and
"probe" beams. The pump beam heats the sample, while the time-
delayed probe beam is used to measure thermally induced changes in
reflectivity at the surface of the sample as a function of time. The experimental data are then fit with an analytical thermal model and properties
such as thermal conductivity and/or interface thermal conductance are
then extracted.

When conducted in the time-domain, these thermoreflectance techniques are often referred to as transient thermoreflectance (TTR) or
time-domain thermoreflectance (TDTR). Several major advantages of
these optical techniques are that (i) they are non-contact and require
little to no sample preparation (other than coating the sample with a
thin metallic film), (ii) they can be used on a variety of solid and liquid materials with a wide range of thermal properties provided that
the sample has a relatively smooth surface, (iii) they can quantitatively
obtain thermal conductivity *and* interface thermal conductance simultaneously, and (iv) the experimental apparatus can easily be modified to

accommodate a variety of other measurements. The primary drawback of TDTR measurements is the initial investment in a femtosecond laser and the associated optical components required for the measurements. This chapter discusses the basic principles and applications of time-domain thermoreflectance and closely related measurement techniques.

13.2. BASIC PRINCIPLES OF TIME-DOMAIN THERMOREFLECTANCE (TDTR)

Picosecond thermoreflectance measurements were first introduced by Paddock and Eesley[7] and by Capinski and coworkers.[8] Numerous recent advances have been made to improve the signal-to-noise ratio and accuracy of the measurements. Additional advances have extended TDTR for quantitative thermal conductivity imaging with $\sim 3\,\mu$m resolution and measurements on liquids and nanoparticle suspensions.

13.2.1. Typical TDTR Experimental Apparatus

A schematic diagram of a typical TDTR system is shown in Fig. 13.1, and the basic principles are described as follows. A Ti:Sapphire mode-locked laser produces a series of sub-picosecond optical pulses in the near infrared (typically ~ 800 nm) at a repetition rate of ~ 80 MHz. These pulses are split into two separate beams, which are commonly referred to as the "pump" and "probe" beams. The pump beam is modulated at a frequency f (typically $f \sim 10$ MHz) with an electro-optic modulator, and the beam is focused with a standard microscope objective lens on the surface of the sample. The $1/e^2$ pump and probe beam diameters can be adjusted down to $\sim 10\,\mu$m, depending on the optical configuration and the objective lens used.

The sample is coated with a thin (~ 50–100 nm) metal film where the reflectivity has at least a modest dependence on temperature. Aluminum is a convenient choice for the metal layer in TDTR as Al has a relatively large reflectivity change with temperature in the near IR ($dR/dT \sim 2 \times 10^{-4}\,\mathrm{K}^{-1}$ at a wavelength of 770 nm)[9,10] and Al is easy to deposit on a wide range of samples. A small fraction of the energy in the pump pulse is absorbed within the first ~ 10 nm of the metal

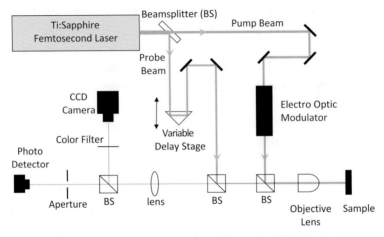

Fig. 13.1   Schematic diagram of a basic TDTR system. A Ti:Sapphire laser produces a
series of subpicosecond laser pulses that are split into pump and probe beams. The pump
beam is modulated with an electro-optic modulator, and the pump beam is then focused on
the sample using an objective lens. The probe beam is also focused on the same spot on
the sample with the objective lens, but the probe follows a different optical path and the
pathlength of the probe is varied through the use of a mechanical delay stage. The pump
beam heats the metallic surface of the sample, and the time-delayed probe beam reflects off
the surface of the sample back to the photodetector. The reflectivity of the metallic coating
on the sample changes as a function of temperature, thus the intensity of the probe beam
that reaches the photodetector gives a measure of the temperature at the sample surface.
The signal from the photodetector is recorded with a lock-in amplifier and the experimental
data is then compared with an analytic thermal model in order to extract thermal properties
from the sample.

creating a nearly instantaneous temperature rise on the order of ~ 5 K
near the surface of the sample. The initial temperature rise and subsequent decay creates a corresponding change in reflectivity of the metal,
and this change in reflectivity is monitored with the probe beam. Wang
et al.[10] recently evaluated dR/dT for 18 metallic elements and found
that Al and Ta have the largest dR/dT at 785 nm while Nb, Re, Ta and
V have the largest thermoreflectance at 1.55 μm.

   The probe beam follows a different optical path to the sample and
this pathlength is adjustable through the use of a mechanical delay stage
equipped with a corner cube retroreflector. A delay stage with 500 mm of
travel translates into ~ 3 ns of delay between the pump and probe beams.
The time-delayed probe beam passes through the same objective lens as
the pump beam, and it is focused on the same spot on the sample. The

probe beam reflects off of the sample, back through the objective lens, and into a high-speed photodetector (Si photodiode). Changes in the intensity of the reflected probe beam appear at the modulation frequency of the pump beam and are measured at the photodiode detector using a radiofrequency (RF) lock-in amplifier. The lock-in amplifier records the in-phase voltage, V_{in}, and out-of-phase voltage, V_{out}, from the photodetector as a function of delay time, t, where $V_f(t) = V_{in}(t) + iV_{out}(t)$, and these data are compared with a thermal model,[9] described in Sec. 13.2.2, in order to extract the thermal properties of interest. Figure 13.2 shows typical TDTR data that has been fit with a thermal model.

The penetration depth of the thermal wave created by the periodic heating in TDTR is controlled by the modulation frequency, f. The penetration depth, d, of the thermal wave can be defined as the distance from the surface at which the amplitude of the temperature oscillation is reduced by a factor of $1/e$.[11] With this definition,

$$d = \sqrt{\frac{k}{\pi C f}},$$ (13.1)

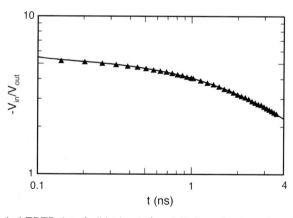

Fig. 13.2   Typical TDTR data (solid triangles) and fit from the thermal model (line). The ratio of the in-phase to out-of-phase voltage $(-V_{in}/V_{out})$ is measured by the lock-in amplifier as a function of the delay time, t, between the arrival of the pump and probe beams at the sample. The thermal conductivity of the sample and the interface thermal conductance between the aluminum film and the sample are used as fitting parameters in the thermal model, and they are adjusted until the model fits the experimental data.

where k and C are the thermal conductivity and volumetric heat capacity of the sample, respectively. Therefore, by changing the modulation frequency of the pump beam, one can preferentially examine different depths and different properties of the sample. The effect of modulation frequency on the measurement sensitivity is discussed in more detail in Sec. 13.2.2.

For certain samples, one also should be concerned with the possible frequency dependence of the thermal conductivity. Previous work[12,13] has shown that semiconductor alloys display a reduction in thermal conductivity at high modulation frequencies. Koh and Cahill[12] speculated that the thermal conductivity measured by TDTR does not include heat carried by phonons that have mean-free paths larger than the thermal penetration depth of the measurement.

In the past decade there have been numerous advances that have made TDTR measurements simpler, more accurate, and more robust than the pioneering thermoreflectance measurements of the mid to late 1980's. Of primary importance has been the rapid advance in the technology of Ti:Sapphire laser systems. Current femtosecond systems are stable with low-noise and are accessible and easy to use, even for researchers with little background in the field of ultra-fast lasers. Other advances include the use of a CCD camera to capture a dark-field image of the sample surface.[1] The CCD camera is placed at the focal length of the lens that is used to focus the reflected probe beam on the photodetector. This image of the sample surface is useful for focusing and aligning the pump and probe beams as well as for ease in navigating to desired regions of the sample.

A significant advance in the analysis of the TDTR data is the use of the out-of-phase signal in addition to the typical analysis of the in-phase signal. Many non-idealities in the measurement, such as defocusing of the delayed probe beam, or changes in the pump-probe overlap at the sample surface, affect both V_{in} and V_{out} in the same manner. Therefore, by analyzing the ratio of V_{in}/V_{out} many of the non-idealities in the system can be minimized or eliminated. Many other additional recent advances are discussed in more detail throughout the chapter.

One interesting side benefit of TDTR measurements is that, in some cases, picosecond ultrasonic data can be used to measure film

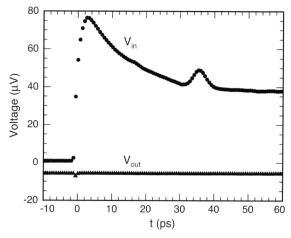

Fig. 13.3 Sample TDTR data that show an acoustic echo at short delay times, *t*. V_{in} and V_{out} are the in-phase and out-of-phase voltages, respectively, recorded at the photodetector. When energy from the pump beam is absorbed at the surface of the metal film, a strain wave is created and this wave reflects off the interface between the aluminum and the sample. When the wave returns to the metal surface it induces a change in reflectivity of the metal as seen here for $t \sim 35$ ps. One can use the round trip time for the echo to appear, along with the speed of sound in aluminum (~ 6420 m/s), to calculate the thickness of the aluminum film. Also note that the out-of-phase voltage remains constant as the delay time crosses zero. This feature is used to correctly set the phase for the measurements, as the phase on the lock-in can be adjusted until V_{out} remains constant across $t = 0$.

thicknesses. When the metal film absorbs the laser energy in the pump beam, a sound wave is launched from the surface of the metal film. This sound wave travels through the metal film and reflects off of the interface between the metal film and the underlying layer. When the wave returns to the surface the change in strain affects the reflectivity of the metal and leaves a peak in the in-phase data, as shown in Fig. 13.3. If the speed of sound is known for the metal film, the round trip time for the acoustic echo can be used to calculate the thickness of the aluminum film. Depending on the acoustic impedance of the layers, multiple echoes, or even echoes from additional layers may be observed.

13.2.1.1.   Exclusion of diffusely scattered light from the pump beam

One significant source of error that researchers need to carefully consider is scattered light from the pump beam that reaches the photodetector.

Since the lock-in amplifier measures all incident light at the modulation frequency that reaches the photodetector, rough samples that diffusely scatter light from the pump beam back to the detector can cause spurious results. Thus, one must take care to either (i) prevent scattered light from the pump beam from reaching the detector, or (ii) use strategies to eliminate the effect of the scattered pump beam that does reach the detector. There are numerous approaches of varying degrees of complexity that can be used to eliminate the effect of scattered light from the pump beam.

(i) For smooth samples that act as specular reflectors, light from the pump beam can sufficiently be excluded from the photodetector through the use of orthogonal polarizations for the pump and probe beams, a polarizing beam splitter, separate optical paths for the pump and probe beams through the objective lens, and an aperture before the detector.

(ii) The relative contributions of the pump and probe beams can be separated out if the probe beam is modulated at low frequencies.[14] In this "double modulation" scheme a mechanical chopper is used to modulate the probe beam at a frequency $\sim$200 Hz. The signal from the photodetector is first fed to the RF lock-in, and the in-phase and out-of-phase outputs from the RF lock-in are then separately sent to two audio frequency lock-ins that use the chopper modulation frequency as the reference.

(iii) A more complicated approach is to use different wavelengths for the pump and probe beams.[15,16] One common technique in a "two-color" experiment is to frequency double the pump or probe beam using a nonlinear crystal such as beta barium borate (BBO), or bismuth borate (BiBO). With a large separation between the pump and probe wavelength, one can use dielectric mirrors and color filters to completely isolate the probe from the pump. This technique can simplify the alignment as the pump and probe beams can be co-aligned, but the nonlinear optics can increase the noise in the pump beam and reduce the overall signal-to-noise ratio. One also needs to consider the possible effects on the sample by using UV light as the pump.

(iv) A relatively simple approach to suppress the scattered light from the pump beam is to use sharp-edged optical filters to split the broadband output from the Ti:Sapphire laser into spectrally distinct pump and probe beams.[17] Here a combination of long-wave pass, short-wave pass, and bandpass filters are used to split the long wavelength portion of the broadband Ti:Sapphire output into the pump beam, while the probe is created from the short wavelength tail of the Ti:Sapphire output. This creates a spectral separation of ~10 nm between the pump and probe, and a short-wave pass filter placed in front of the photodetector can reduce the scattered light from the pump by a factor of ~1000.

13.2.1.2.   *Additional experimental details*

Any number of minor modifications to the experimental apparatus can be beneficial for obtaining accurate and repeatable data for a given application. Similarly, there are several pitfalls that one should be careful to avoid when assembling a TDTR system and performing measurements. Several such details are listed below.

(i) Depending on the laser system, an optical isolator may be required to keep reflections of the laser pulses off of optical components from returning to the laser cavity as these reflections can disturb the mode-locking ability of the laser.

(ii) Placing an inductor in series between the photodetector and the RF lock-in to create a passive resonant circuit will significantly enhance the signal-to-noise ratio.[18] The inductor and the stray capacitance of the photodetector create a series $L-C$ circuit that can increase the signal by a factor of ~10. The inductor should be selected to maximize the signal at the desired modulation frequency.

(iii) One should note that the delay stage can be used to adjust the pathlength of the pump beam rather than the pathlength of the probe beam. In other words, the arrival of the pump beam at the sample can be advanced relative to the probe, or the arrival of the probe beam can be delayed relative to the pump. The "advanced pump" case does introduce an additional phase shift into the data analysis,[9] but this is easy to account for the thermal model. Experimentally, the advantage of the advanced pump case is that the probe beam is stationary throughout the measurement and it does not wander on the photodetector. However, with modest care in alignment, the movement of the probe beam can be reduced such that it is of little practical concern. The disadvantage of the advanced pump case is that as the delay stage moves, the pump beam may wander slightly which can alter the amount of scattered light from the pump beam that reaches the photodetector. This scattered light can be accounted for by subtracting off a background scan (i.e. a scan with the probe beam blocked) from the raw data. In practice, either method is suitable and the choice may be left to the layout of the optics on the table or arrangement of the optics to accommodate other experiments.

(iv) The divergence of the laser system plays an important role in the spot size of the laser at the sample. For certain laser systems and optical configurations, some researchers may find it useful to use additional optics, such as a Keplerian beam expander, to control the spot size and the divergence. Also note that large divergence will create a large change in the spot size of the delayed beam as the distance on the delay stage changes. However, this change in the diameter of the beam can be accounted for in the thermal model.[13]

(v) One potentially large source of error that requires careful attention is the phase of the reference channel on the lock-in.[9] A convenient method for correctly setting the phase is to examine the out-of-phase voltage, $V_{out}(t)$, near $t = 0$, since this voltage should remain constant as the delay time crosses from negative to positive delay (see Fig. 13.3). Small errors in the phase, ε, can be corrected after the raw data is acquired by multiplying $V_f(t) = V_{in}(t) + iV_{out}(t)$ by a small phase correction factor of $(1 - i\varepsilon)$. Errors in setting the phase can be the dominant source of uncertainty for measurements ~ 1 MHz or less.

(vi) The magnitude of the signal is determined by the *product* of the pump and probe beam energies. Thus the pump and probe beams do not need to be split in a 95/5 manner as is typical for other pump-probe measurements.

(vii) The TDTR experiments can be simplified and automated through the use of software, such as LabVIEW®, for instrument control and data acquisition. With a simple program to control the motion of the delay stage and to record the data from the lock-in amplifier, a 4 ns scan on a sample can be performed in a matter of minutes. A motorized sample stage can also be used for automated measurements on different regions of the sample.

13.2.2. Data Analysis

The thermal conductivity of the sample and the interface thermal conductance are extracted by comparing the experimental data to a theoretical model (see Fig. 13.2).[9] Typically, the interface conductance between two layers and the thermal conductivity of a single layer can be extracted simultaneously by using these two properties as fitting parameters in the model. The analytic model starts with the solution to the heat equation for a semi-infinite solid that is heated at the surface by a periodic point source as given by Carslaw and Jaeger.[19] With some manipulation, the solution to the heat equation can be generalized for a layered geometry.[20] The final development of the full model is not entirely straightforward, but the model is described in detail by Cahill.[9] Further refinements of the thermal model include accounting for changes in the diameter of the beam that is on the optical delay line,[13] and accounting for anisotropic thermal conductivities in the layered media.[21] The general elements of the model are described in this section.

The thermal model includes the thermal conductivity, volumetric heat capacity, and thickness of each layer in the sample along with the laser repetition rate, the modulation frequency, and the pump and probe beam diameters. The general model considers radial heat flow, however, since the penetration depth of the thermal wave is generally much smaller than the $1/e^2$ radius of the pump and probe beams, the heat flow is predominantly one dimensional and the measured thermal conductivity is generally the conductivity in the direction normal to the surface.

For thin aluminum layers, the thermal conductivity is reduced slightly from the bulk value. The thermal conductivity of the aluminum layer is typically determined from the Wiedemann-Franz law, which relates the thermal conductivity, k, to the electrical conductivity, σ, of a metal through $k = L\sigma T$, where L is a proportionality constant called the Lorenz number, and T is absolute temperature. Thus with a simple four point probe measurement of the electrical conductivity of the aluminum film, one can quickly estimate the thermal conductivity of the aluminum.

The thickness of the aluminum can be determined from the picosecond acoustics measurements, cross-sectional transmission electron microscopy analysis, or Rutherford backscattering spectroscopy. The volumetric heat capacity for all layers is generally assumed to be equal to bulk values.

The thermal resistance created by an interface is included in the model by including a thin (e.g. 1 nm) layer to represent the interface. The ratio of the thermal conductivity to the thickness of this "artificial" layer gives the interface conductance, G, and this ratio is frequently used as a fitting parameter to the experimental data. Since the interface is represented as a thin layer, the heat capacity of this layer has no effect on the model.

The properties of additional layers and interfaces are entered in the model in a similar fashion where the thermal conductivity of one layer is unknown and this conductivity is used as a fitting parameter in the model. For multilayer samples where multiple layers may have unknown thermal conductivity, an additional "reference" sample consisting of a single layer is frequently prepared so that the conductivity of that layer can be measured separately.

13.2.2.1. Sensitivity analysis

With a simple sensitivity analysis,[22] one can determine the viability of measurements on a particular sample, optimize the design of a sample, determine the optimum frequency at which to conduct an experiment, and gain a quantitative understanding of the importance of certain parameters involved in the experiment. For TDTR measurements, a sensitivity parameter, S_x can be defined as the sensitivity of the model to the logarithmic derivative of the ratio V_{in}/V_{out} with respect to a particular parameter, x, i.e.

$$S_x = \frac{d \ln(-V_{in}/V_{out})}{d \ln x}. \qquad (13.2)$$

Take, for example, the simple case of a film with $k_{film} = 10\,\mathrm{W\,m^{-1}\,K^{-1}}$ and $C_{film} = 1.5\,\mathrm{W\,cm^{-3}\,K^{-1}}$ deposited on a silicon substrate and coated with an aluminum film with thickness of $100\,\mathrm{nm}$, $k_{Al} = 200\,\mathrm{W\,m^{-1}\,K^{-1}}$, and $C_{Al} = 2.4\,\mathrm{J\,cm^{-3}\,K^{-1}}$. If the film could be grown to thicknesses between $100\,\mathrm{nm}$ and $1\,\mu\mathrm{m}$, a sensitivity analysis as shown in Fig. 13.4 is useful in understanding a few basic trends.

First, for the $100\,\mathrm{nm}$ film in Fig. 13.4(a), we see that accurate determination of the aluminum film thickness is critical and that the measurement is more sensitive to the conductance at the film/substrate interface, G_2, than the conductance at the Al/film interface, G_1. However, for a $1\,\mu\mathrm{m}$ thick film measured at the same frequency (Fig. 13.4(b)), the penetration depth of the thermal wave is now confined to the film and the measurement is not at all sensitive to G_2 or the thermal conductivity of the substrate. For this configuration, we also notice a typical feature in TDTR measurements in that the experiment is sensitive to the thermal conductivity of the film, k_{film}, at short, $t < 1\,\mathrm{ns}$, delay times, while the sensitivity to G_1 increases at longer delay times. By fitting the model to the data at short or long delay times one can improve the accuracy of determining either k_{film} or G_1.

Figures 13.5(a) and (b) show the sensitivities for the same $100\,\mathrm{nm}$ and $1\,\mu\mathrm{m}$ thick films as a function of modulation frequencies between $250\,\mathrm{kHz}$ and $20\,\mathrm{MHz}$ with the delay time fixed at $0.2\,\mathrm{ns}$. Here the penetration depth of the thermal wave increases at low frequency and the influence of the substrate becomes important. For the thin film in Fig. 13.5(a), we see that at low frequencies the measurement is sensitive primarily to the thermal conductivity of the substrate and the thickness of the metal film. However, as the penetration depth of the thermal wave decreases at higher frequencies the importance of the substrate thermal conductivity decreases and the influence of the film and the interfaces increases. For the $1\,\mu\mathrm{m}$ thick film in Fig. 13.5(b), several interesting trends are apparent. Most notable is that the sensitivity to the thickness of the film crosses zero at $\sim4\,\mathrm{MHz}$, while there is still significant sensitivity to the thermal conductivity of the film. Thus $4\,\mathrm{MHz}$ would be an optimal frequency for measurements of the thermal conductivity of the film.

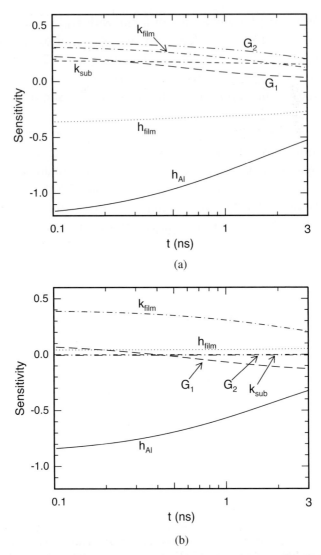

Fig. 13.4 Sensitivity analyses as a function of delay time for a simple case with an aluminum coated film on a silicon substrate with a modulation frequency of $f = 10\,\text{MHz}$. In (a) the thickness of the film, h_f, is 100 nm and in (b) $h_f = 1000\,\text{nm}$. The aluminum film has a thickness of $h_{Al} = 100\,\text{nm}$, volumetric heat capacity of $C_{Al} = 2.4\,\text{J cm}^{-3}\,\text{K}^{-1}$ and thermal conductivity of $k_{Al} = 2\,\text{W cm}^{-1}\,\text{K}^{-1}$. The interfaces between the aluminum and the film, G_1, and the film and the substrate, G_2, both have a thermal conductance of $100\,\text{MW m}^{-2}$ K^{-1}. The film and substrate properties are $k_f = 0.1\,\text{W cm}^{-1}\,\text{K}^{-1}$, $C_f = 1.5\,\text{J cm}^{-3}\,\text{K}^{-1}$, $k_{sub} = 1.48\,\text{W cm}^{-1}\,\text{K}^{-1}$, and $C_{sub} = 1.64\,\text{J cm}^{-3}\,\text{K}^{-1}$, respectively.

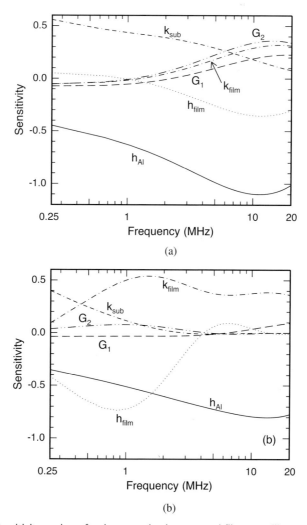

Fig. 13.5   Sensitivity analyses for the same aluminum coated film on a silicon substrate as in Fig. 13.4. Again in (a) the thickness of the film, h_f, is 100 nm and in (b) $h_f = 1000$ nm. Here the modulation frequency varies from 250 kHz to 20 MHz and the sensitivities are presented at a delay time of $t = 0.2$ ns.

Readers familiar with the 3ω method for determining thermal conductivity of thin films are encouraged to consult a recent study by Koh *et al.*[12] where they used both TDTR and 3ω to examine several samples of $(In_{0.52}Al_{0.48})_x(In_{0.53}Ga_{0.47})_{1-x}$ As alloy layers with embedded ErAs nanoparticles. They found that the measurements agreed to within

experimental uncertainty at frequencies around 1 MHz and they discussed accuracy, limitations, and uncertainties of both methods.

13.3.   EXAMPLES OF TDTR APPLICATIONS

Time domain thermoreflectance is a robust measurement technique
that has been used to measure a wide range of solid and liquid samples in the past decade. Additionally, with some minor modifications,
the apparatus used for TDTR can be readily adjusted for a variety of
related measurements. A brief sampling of some noteworthy applications are discussed below, and many more examples can be found in the
literature.

TDTR has been used for measurements of the interface thermal
conductance of highly dissimilar materials[23] as well as metal-metal interfaces[22] that span a range of $8 < G < 4000$ MW m^{-2} K^{-1}. Costescu
et al.[24] and Chiritescu et al.[25] have measured ultra-low thermal conductivity in W/Al$_2$O$_3$ nanolaminates ($k \sim 0.6$ W m^{-1} K^{-1}) and layered
WSe$_2$ crystals ($k \sim 0.05$ W m^{-1} K^{-1}), respectively. Persson et al.[11]
examined the thermal conductance of InAs nanowire composites with
TDTR and a finite element method analysis of the complex compos-
ite structure. Shukla et al.[26] used TDTR to examine nanostructured
composites of nickel nanoparticles embedded in a yttrium stabilized zirconia matrix and extracted interface thermal conductance for the buried
nanoparticles. The same group also measured interface thermal conductance for amorphous and crystalline metallic alloys and buried interfaces
by varying measurement frequency.[27]

With only the slightest of modifications, a TDTR system can be
used to make rapid and quantitative thermal conductivity images of
materials with a resolution of ~ 3 μm.[14,28–30] As mentioned previously,
at short delay times the ratio V_{in}/V_{out} often has a stronger dependence
on k than G. In these imaging experiments that dependence is taken
advantage of by fixing the delay stage at $t \sim 100$ ps and raster scanning
the sample using a motorized stage. The ratio V_{in}/V_{out} is recorded at
each point and fit to the same thermal model. However, since the scans
are performed at a single delay time, the accuracy of these measurements
is worse than standard TDTR measurements due to uncertainties in the

phase and G of the Al/sample interface. Images with 100×100 pixels can be obtained is less than an hour. This method has been used to obtain thermal conductivity images of a diffusion multiple,[14] measurements of the depth profile of thermal conductivity in thermal barrier coatings,[29] and images of nickel solid solutions.[28]

TDTR has also been used by Ge *et al.*[31] to measure the thermal conductance across planar interfaces between water and solids that were functionalized with hydrophobic and hydrophilic self-assembled mono-layers (SAMs). A schematic diagram of the sample structure used for measuring thermal transport across the aluminum-water interface is shown in Fig. 13.6. In these measurements, the pump and probe beams first pass through the anti-reflective SiO_2 layer, the sapphire substrate, and the polyimide insulating layer, before being absorbed by the Ti and

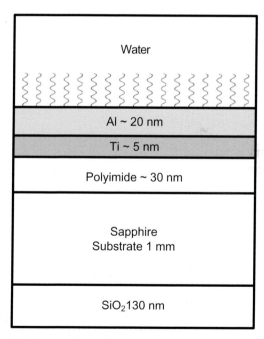

Fig. 13.6   Schematic diagram of the sample structure used by Ge *et al.*[31] to examine the thermal conductance between an aluminum-water interface where the aluminum was coated with a self assembled monolayer. Here the pump and probe beams are incident through the transparent SiO_2, sapphire, and polyimide layers. The laser energy is absorbed by the Ti and Al layers and a bidirectional heat flow model is used to extract the thermal conductance across the Al-water interface.

Al layers. The heat then diffuses through the SAM into the water and back in the opposite direction towards the sapphire substrate. The poly-imide layer is used as insulation in order to reduce the backflow of heat to the sapphire and increase the measurement sensitivity to the solid-liquid interface. This experimental arrangement requires that the thermal model account for the bi-directional heat flow.[9]

Schmidt *et al.*[16] have also used a slightly modified TDTR approach for measuring thermal conductivity of liquids that does not depend on the optical properties of the liquid and can measure samples as small as a single drop. Here an aluminum film is deposited on a glass slide and a small volume of liquid is contained on the surface of the aluminum. As in the previous example, the pump and probe beams pass through the transparent substrate and are incident on the aluminum film. The heat flow is again bi-directional, and here the signal phase is fit with a thermal model in order to extract the thermal conductivity of the liquid.

In addition to their work with liquids, Schmidt *et al.*[21] also examined the effects of pulse accumulation on radial heat conduction and explained how TDTR could be used to probe in-plane thermal conductivity in addition to typical cross-plane measurements. They used this technique to extract cross-plane and in-plane thermal conductivities for highly ordered pyrolytic graphite.

Two important features of TDTR measurements are that the measurements are non-contact and do not require large samples. Therefore, one can perform measurements on small samples in any number of environments, where the only significant restriction is optical access to the sample. For example, measurements on samples with diameters ~100 μm at high-pressure[32] using a diamond anvil cell are possible, and measurements at high or low temperatures can be accommodated in a cryostat or oven. The greatest limitation to the measurements is the physical and chemical stability of the sample and metal film.

13.3.1. *Measurement Techniques Closely Related to TDTR*

The experimental apparatus used for TDTR can also be used to perform measurements in the frequency domain rather than the time domain.[33–35] For frequency domain thermoreflectance (FDTR), the delay stage is fixed at a certain delay time, and the thermoreflectance signals are measured at various modulation frequencies allowing for the examination of a range of thermal penetration depths in a single experiment. Since the delay time is fixed in FDTR, this technique has the advantage of avoiding errors that are introduced in TDTR due the motion of the delay stage. Schmidt et al.[33] used this technique to extract heat capacity, in-plane and cross-plane thermal conductivity, and inter-face thermal conductance for several materials. They also point out that a modified FDTR technique can be implemented with a continuous-wave laser, leading to a simpler and less expensive measurement system.

The TDTR system can also be modified for an Ångström method for measuring the in-plane thermal diffusivity for samples with the proper geometry.[36] In general, the Ångström technique is a frequency domain measurement where one location of a sample is heated periodically and the thermal decay and phase lag are measured as a function of distance from the heat source. When adapting the standard TDTR system, the pump beam again serves as the heat source and the probe beam measures the resulting thermal decay, except now the pump and probe beams are separated on the sample instead of being co-aligned on the same location.

A technique that is closely related to TDTR is called transient absorption. Transient absorption measurements operate in a similar fashion to TDTR except that they rely on small changes in absorption, rather than reflection, with temperature. These measurements are particularly useful for examining the interface thermal conductance for nanoparticles suspended in a liquid. Here the pump beam passes through the suspension and a small fraction of the pump energy is absorbed by the nanoparticles. The time-delayed probe beam passes through the same region of the suspension and is then focused on the photodetector. If the absorptivity of the nanoparticles depends on temperature, then the fraction of the probe beam energy that reaches the detector is related to the temperature of the nanoparticles. Thus, the transient absorption signal recorded by the lock-in gives a measure of the temperature decay of the nanoparticles as a function of time. For sufficiently small nanoparticles where the thermal decay is controlled by the interface, the interface conductance can be determined from[37]

$$G = \frac{rC_p}{3\tau} \qquad (13.3)$$

where τ is the characteristic time for the cooling of the nanoparticle, r is the nanoparticle radius, and C_p is the heat capacity per unit volume of the nanoparticle. A typical TDTR configuration can easily be modified in a matter of minutes to accommodate transient absorption experiments by simply swapping samples and adjusting the sample stage and lens behind the stage.

Transient absorption was used to perform the first measurements of the interface thermal conductance between single-walled carbon nanotubes and a surfactant,[38] and to examine the interface conductance for metallic[37,39] and Au-core polymer-shell[40] nanoparticles suspended in various solvents. Schmidt et al.[41] used transient absorption to study heat transfer through the interfaces of ligand passivated gold nanorods in suspension. This technique has also been used to examine the vibrational decay of higher order fullerenes suspended in a variety of solvents.[42] Care should be taken when examining nanoparticles, such as single-walled carbon nanotubes, as the absorptivity, and the change in absorptivity with temperature, can have a complicated dependence on wavelength in the near IR.[38]

While not directly related to TDTR, readers interested in other optical techniques for thermal characterization of nanoparticle suspensions and molecular chains are directed to several recent papers from the Cahill group at the University of Illinois at Urbana-Champaign.[43-48] Putnam et al. developed a micrometer scale ac beam deflection technique to measure thermodiffusion in liquids[43-45] and thermal conductivity of nanoparticle suspensions.[46] This technique is based on the deflection of a laser beam created by a thermally induced gradient in the index of refraction in the fluid. Wang et al.[47,48] developed an ultrafast flash thermal conductance apparatus where a sub-picosecond laser pulse was used to heat a metal surface and heat flow from the metal surface through molecules is detected using vibrational sum-frequency generation (SFG) spectroscopy.

An extension of TDTR that is used for measuring the coefficient of thermal expansion is called time-domain probe beam deflection[49,50]

(TD-PBD). In TD-PBD, the pump and probe beams are separated by a few microns on the surface of the sample. The modulated pump beam heats the surface of the sample and thermal expansion and elastic deformation of the sample create vertical displacements of the surface. These displacements produce small beam deflections of the nearby probe beam. The probe beam deflections are monitored by a split Si photodiode and an RF lock-in. As in TDTR, the ratio of $V_{in}(t)/V_{out}(t)$ over short delay times, $-20\,\text{ps} < t < 300\,\text{ps}$, is compared with a quantitative thermal model and the coefficient of thermal expansion is extracted.

Another use for a modified pump-probe laser system is for the examination of melting and solidification of materials.[51–53] An amplified femtosecond laser is used to melt a thin layer (10–20 nm) of the sample and a time delayed probe beam measures the thickness of the liquid layer as a function of time with optical third-order harmonic generation of light.

13.4. SUMMARY AND OUTLOOK

This chapter has outlined the basic principles of time domain thermoreflectance measurement systems. Since TDTR is a non-contact measurement technique that requires little sample preparation besides a thin metallic coating, this technique has become a robust and powerful tool for thermal characterization of various types of nanostructured materials, including thin films, composites, layered media, nanowire arrays, and colloidal suspensions of nanoparticles. Recent extensions of the basic TDTR system, including transient absorption, probe beam deflection, and frequency domain thermoreflectance have further increased the utility of TDTR based experimental systems.

In the near future, the need for thermal characterization of nanostructured materials will undoubtedly increase as the performance of many microelectronic, thermoelectric, and photonic devices depends critically on the thermal properties of the thin films, nanowires, and composites within these devices. TDTR will likely play an important role in characterizing new nanostructured materials as commercial laser systems continue to deliver superior performance coupled with decreasing price and user complexity each year, thus making these pump-probe type of measurement systems accessible to more researchers with little background in ultrafast lasers. Finally, as more research groups develop TDTR systems and become accustomed to pump-probe measurements for thermal characterization, there will also be a corresponding increase in the development and refinement of systems and techniques related to TDTR for specialized thermal measurements of complex nanostructures.

Acknowledgments

STH acknowledges support from the US National Science Foundation under Grant No. CBET 0547122 and thanks Harikrishna for assistance with some of the figures in this chapter.

REFERENCES

1. D. G. Cahill, K. Goodson, A. Majumdar, *J. Heat Transfer-Transactions ASME* **124**, 223 (2002).
2. D. G. Cahill, W. K. Ford, K. E. Goodson, G. D. Mahan, A. Majumdar, H. J. Maris, R. Merlin, S. R. Phillpot, *J. Appl. Phys.* **93**, 793 (2003).
3. D. G. Cahill, *Rev. Sci. Instrum.* **61**, 802 (1990).
4. A. Majumdar, *Ann. Rev. Mat. Sci.* **29**, 505 (1999).
5. R. M. Costescu, M. A. Wall, D. G. Cahill, *Phys. Rev. B* **67**, 054302 (2003).
6. A. N. Smith, J. L. Hostetler, P. M. Norris, *Microscale Thermophysical Engineering* **4**, 51 (2000).
7. C. A. Paddock, G. L. Eesley, *J. Appl. Phys.* **60**, 285 (1986).
8. W. S. Capinski, H. J. Maris, T. Ruf, M. Cardona, K. Ploog, D. S. Katzer, *Phys. Rev. B* **59**, 8105 (1999).
9. D. G. Cahill, *Rev. Sci. Instrum.* **75**, 5119 (2004).
10. Y. Wang, J. Park, Y. K. Koh, D. G. Cahill, *J. Appl. Phys.* **108**, 043507 (2010).
11. A. I. Persson, Y. K. Koh, D. G. Cahill, L. Samuelson, H. Linke, *Nano Lett.* **9**, 4484 (2009).
12. Y. K. Koh, S. L. Singer, W. Kim, J. M. O. Zide, H. Lu, D. G. Cahill, A. Majumdar, A. C. Gossard, *J. Appl. Phys.* **105**, 054303 (2009).
13. Y. K. Koh, D. G. Cahill, *Phys. Rev. B* **76**, 075207 (2007).
14. S. Huxtable, D. G. Cahill, V. Fauconnier, J. O. White, J. C. Zhao, *Nat. Mat.* **3**, 298 (2004).
15. T. Asahi, A. Furube, H. Fukumura, M. Ichikawa, H. Masuhara, *Rev. Sci. Instrum.* **69**, 361 (1998).
16. A. Schmidt, M. Chiesa, X. Y. Chen, G. Chen, *Rev. Sci. Instrum.* **79**, 064902 (2008).
17. K. Kang, Y. K. Koh, C. Chiritescu, X. Zheng, D. G. Cahill, *Rev. Sci. Instrum.* **79**, 114901 (2008).

18. K. E. O'Hara, X. Y. Hu, D. G. Cahill, *J. Appl. Phys.* **90**, 4852 (2001).
19. H. S. Carslaw, J. C. Jaeger, *Conduction of Heat in Solids*, Oxford University Press, New York (1959).
20. A. Feldman, *High Temperatures-High Pressures* **31**, 293 (1999).
21. A. J. Schmidt, X. Y. Chen, G. Chen, *Rev. Sci. Instrum.* **79**, 114902 (2008).
22. B. C. Gundrum, D. G. Cahill, R. S. Averback, *Phys. Rev. B* **72**, 245426 (2005).
23. H. K. Lyeo, D. G. Cahill, *Phys. Rev. B* **73**, 144301 (2006).
24. R. M. Costescu, D. G. Cahill, F. H. Fabreguette, Z. A. Sechrist, S. M. George, *Science* **303**, 989 (2004).
25. C. Chiritescu, D. G. Cahill, N. Nguyen, D. Johnson, A. Bodapati, P. Keblinski, P. Zschack, *Science* **315**, 351 (2007).
26. N. C. Shukla, H. H. Liao, J. T. Abiade, M. Murayama, D. Kumar, S. T. Huxtable, *Appl. Phys. Lett.* **94**, 151913 (2009).
27. N. C. Shukla, H. H. Liao, J. T. Abiade, F. X. Liu, P. K. Liaw, S. T. Huxtable, *Appl. Phys. Lett.* **94**, 081912 (2009).
28. X. Zheng, D. G. Cahill, P. Krasnochtchekov, R. S. Averback, J. C. Zhao, *Acta Materialia* **55**, 5177 (2007).
29. X. Zheng, D. G. Cahill, J. C. Zhao, *Adv. Eng. Mat.* **7**, 622 (2005).
30. E. Lopez-Honorato, C. Chiritescu, P. Xiao, D. G. Cahill, G. Marsh, T. J. Abram, *J. Nuc. Mat.* **378**, 35 (2008).
31. Z. B. Ge, D. G. Cahill, P. V. Braun, *Phys. Rev. Lett.* **96**, 186101 (2006).
32. W. P. Hsieh, B. Chen, J. Li, P. Keblinski, D. G. Cahill, *Phys. Rev. B* **80**, 180302 (2009).
33. A. J. Schmidt, R. Cheaito, M. Chiesa, *Rev. Sci. Instrum.* **80**, 094901 (2009).
34. A. J. Schmidt, R. Cheaito, M. Chiesa, *J. Appl. Phys.* **107**, 024908 (2010).
35. J. Zhu, D. Tang, W. Wang, J. Liu, R. Yang, in *14th International Heat Transfer Conference (IHTC14)*, Washington, DC, USA (2010).
36. S. T. Huxtable, D. G. Cahill, L. M. Phinney, *J. Appl. Phys.* **95**, 2102 (2004).
37. O. M. Wilson, X. Y. Hu, D. G. Cahill, P. V. Braun, *Phys. Rev. B* **66**, 224301 (2002).
38. S. T. Huxtable, D. G. Cahill, S. Shenogin, L. P. Xue, R. Ozisik, P. Barone, M. Usrey, M. S. Strano, G. Siddons, M. Shim, P. Keblinski, *Nat. Mat.* **2**, 731 (2003).
39. Z. B. Ge, D. G. Cahill, P. V. Braun, *J. Phys. Chem. B* **108**, 18870 (2004).
40. Z. B. Ge, Y. J. Kang, T. A. Taton, P. V. Braun, D. G. Cahill, *Nano Lett.* **5**, 531 (2005).
41. A. J. Schmidt, J. D. Alper, M. Chiesa, G. Chen, S. K. Das, K. Hamad-Schifferli, *J. Phys. Chem. C* **112**, 13320 (2008).
42. S. T. Huxtable, D. G. Cahill, S. Shenogin, P. Keblinski, *Chem. Phys. Lett.* **407**, 129 (2005).
43. S. A. Putnam, D. G. Cahill, *Rev. Sci. Instrum.* **75**, 2368 (2004).
44. S. A. Putnam, D. G. Cahill, *Langmuir* **21**, 5317 (2005).
45. S. A. Putnam, D. G. Cahill, G. C. L. Wong, *Langmuir* **23**, 9221 (2007).
46. S. A. Putnam, D. G. Cahill, P. V. Braun, Z. B. Ge, R. G. Shimmin, *J. Appl. Phys.* **99**, 084308 (2006).

47. Z. H. Wang, D. G. Cahill, J. A. Carter, Y. K. Koh, A. Lagutchev, N. H. Seong, D. D. Dlott, *Chem. Phys.* **350**, 31 (2008).
48. Z. H. Wang, J. A. Carter, A. Lagutchev, Y. K. Koh, N. H. Seong, D. G. Cahill, D. D. Dlott, *Science* **317**, 787 (2007).
49. X. Zheng, D. G. Cahill, R. Weaver, J. C. Zhao, *J. Appl. Phys.* **104**, 073509 (2008).
50. D. G. Cahill, X. Zheng, J. C. Zhao, *J. Thermal Stresses* **33**, 9 (2010).
51. W. L. Chan, R. S. Averback, D. G. Cahill, A. Lagoutchev, *Phys. Rev. B* **78**, 214107 (2008).
52. W. L. Chan, R. S. Averback, D. G. Cahill, Y. Ashkenazy, *Phys. Rev. Lett.* **102**, 095701 (2009).
53. B. C. Gundrum, R. S. Averback, D. G. Cahill, *Appl. Phys. Lett.* **91**, 011906 (2007).

INDEX